Virtual Power Plant Solution for Future Smart Energy Communities

Virtual Power Plant Solution for Future Smart Energy Communities

Edited by
Ehsan Heydarian-Forushani
Hassan Haes Alhelou
Seifeddine Ben Elghali

CRC Press is an imprint of the
Taylor & Francis Group, an **informa** business

MATLAB® is a trademark of The MathWorks, Inc. and is used with permission. The MathWorks does not warrant the accuracy of the text or exercises in this book. This book's use or discussion of MATLAB® software or related products does not constitute endorsement or sponsorship by The MathWorks of a particular pedagogical approach or particular use of the MATLAB® software.

First edition published 2023
by CRC Press
6000 Broken Sound Parkway NW, Suite 300, Boca Raton, FL 33487-2742

and by CRC Press
4 Park Square, Milton Park, Abingdon, Oxon, OX14 4RN

CRC Press is an imprint of Taylor & Francis Group, LLC

© 2023 selection and editorial matter, Ehsan Heydarian-Forushani, Hassan Haes Alhelou and Seifeddine Ben Elghali; individual chapters, the contributors

Reasonable efforts have been made to publish reliable data and information, but the author and publisher cannot assume responsibility for the validity of all materials or the consequences of their use. The authors and publishers have attempted to trace the copyright holders of all material reproduced in this publication and apologize to copyright holders if permission to publish in this form has not been obtained. If any copyright material has not been acknowledged please write and let us know so we may rectify in any future reprint.

With the exception of Chapters 5 and 6, no part of this book may be reprinted or reproduced or utilised in any form or by any electronic, mechanical, or other means, now known or hereafter invented, including photocopying and recording, or in any information storage or retrieval system, without permission in writing from the publishers.

Chapters 5 and 6 of this book is available for free in PDF format as Open Access from the individual product page at www.taylorfrancis.com. It has been made available under a Creative Commons Attribution-Non Commercial-No Derivatives 4.0 license.

For permission to photocopy or use material electronically from this work, access www.copyright.com or contact the Copyright Clearance Center, Inc. (CCC), 222 Rosewood Drive, Danvers, MA 01923, 978-750-8400. For works that are not available on CCC please contact mpkbookspermissions@tandf.co.uk

Trademark notice: Product or corporate names may be trademarks or registered trademarks and are used only for identification and explanation without intent to infringe.

ISBN: 978-1-032-18978-9 (hbk)
ISBN: 978-1-032-18979-6 (pbk)
ISBN: 978-1-003-25720-2 (ebk)

DOI: 10.1201/9781003257202

Typeset in Times
by KnowledgeWorks Global Ltd.

Contents

Preface...vii

Editors..ix

Contributors xi

Introduction..1

Chapter 1 Practical Implementation of VPP in the Real World Based on
Emerging Technologies 7

M. Lakshmi Swarupa, G. Sree Lakshmi, and K. Shashidhar Reddy

Chapter 2 Virtual Power Plant with Demand Response Control in
Aggregated Distributed Energy Resources of Microgrid 39

Elutunji Buraimoh and Innocent E. Davidson

Chapter 3 Exploiting the Flexibility Value of Virtual Power Plants through
Market Participation in Smart Energy Communities..........................55

*Georgios Skaltsis, Stylianos Zikos, Elpiniki Makri, Christos
Timplalexis, Dimosthenis Ioannidis, and Dimitrios Tzovaras*

Chapter 4 Renewable Energy Community VPP Concept Design and
Modelling for Sustainable Islands..79

R. Garner, G. Jansen, and Z. Dehouche

Chapter 5 A Comprehensive Smart Energy Management Strategy for
TVPP, CVPP, and Energy Communities...103

Ehsan Heydarian-Forushani and Seifeddine Ben Elghali

Chapter 6 Virtual Energy Storage Systems for Virtual Power Plants 119

Saif S. Sami, Yue Zhou, Meysam Qadrdan, and Jianzhong Wu

Chapter 7 Centralized and Decentralized Optimization Approaches for
Energy Management within the VPP...145

Nikos Bogonikolos, Entrit Metai, and Konstantinos Tsiomos

v

Chapter 8 Decision-Making Frameworks for Virtual Power
Plant Aggregators in Wholesale Energy and
Ancillary Service Markets..155

S. Bahramara and P. Sheikhahmadi

Chapter 9 Decentralized Energy Management System within VPP 171

K. Shanti Swarup and P. M. Naina

Chapter 10 Distributed Synchronism System Based on TSN and PTP
for Virtual Power Plant..191

*V. Pallares-Lopez, I. M. Moreno-Garcia, R. Real-Calvo,
V. Arenas-Ramos, M. Gonzalez-Redondo, and I. Santiago*

Chapter 11 Complementarity and Flexibility Indexes of
an Interoperable VPP...209

Habib Nasser and Dah Diarra

Chapter 12 Experience of Building a Virtual Power Plant
of the Island of El Hierro 221

*Oleksandr Novykh, Juan Albino Méndez Pérez,
Benjamín González-Díaz, Jose Francisco Gomez Gonzalez,
Igor Sviridenko, and Dmitry Vilenchik*

Index...245

Preface

Virtual power plant's (VPP) goal is to accelerate the transition toward smart and green energy by facilitating the integration of renewable energy sources (RESs). It also aims to help power grids exploit energy efficiency potential and innovative storage approaches, foster the active participation of citizens, and become self-sufficient in energy. All these, while reducing costs, prevent greenhouse gas (GHG) emissions and reliance on fossil fuels to generate power. Another goal is to create new intelligent businesses, growth, and local skilled jobs.

In order to achieve the previous objectives, VPP needs to use disruptive solutions based on forecasting tools, artificial intelligence (AI), Internet of Things (IoT), distributed ledger technology (DLT), and different optimization and control strategies to revolutionize the existing VPP and build smart energy communities. Based on aggregation and smart management of distributed energy resources (DERs), VPP could increase the flexibility and profitability of energy systems while providing novel services. VPP could also enhance the demand-response (DR) capability of consumers by understanding their behaviors and promoting self-consumption.

This book provides a general overview of VPPs as a key technology in future energy communities and active distribution and transmission networks for managing DERs, providing local and global services, and facilitating market participation of small-scale managing DERs and prosumers. The book also aims at describing some practical solutions, business models, and novel architectures for the implementation of virtual power plants in the real world.

Editors

Ehsan Heydarian-Forushani is an assistant professor in the Department of Electrical and Computer Engineering at Qom University of Technology, Qom, Iran. He received his PhD degree in electrical engineering from the Isfahan University of Technology, Isfahan, Iran, in 2017. He was a visiting researcher at the University of Salerno, Salerno, Italy, in 2016–2017. Also, he has received his first postdoc from Isfahan University of Technology, Isfahan, Iran, in 2019. As an industrial experience, he was working in Esfahan Electricity Power Distribution Company (EEPDC) from 2018 to 2021. From February 2021 to January 2022, he was working as a postdoctoral research fellow in Laboratory of Information & Systems (LIS-UMR CNRS 7020), Aix-Marseille University, Marseille, France, on the 7.2-million-euro VPP4ISLANDS ("Virtual Power Plant for Interoperable and Smart isLANDS"), one of the European Union's Horizon 2020 research and innovation programs. His main research interests include power system flexibility, active distribution networks, renewables integration, demand response, smart grids, and electricity market.

Hassan Haes Alhelou is currently working in the Department of Electrical and Computer Systems Engineering, Monash University, Clayton, Victoria, Australia. He was in the School of Electrical and Electronic Engineering, University College Dublin, Dublin, Ireland, and with Sultan Qaboos University, Oman. He is also a professor at Tishreen University, Lattakia, Syria. He is included in the 2018/2019 Publons list of the top 1% best reviewer and researchers in the field of engineering in the world. He was the recipient of the Outstanding Reviewer Award from many journals, e.g., Energy Conversion and Management (ECM), ISA Transactions, and Applied Energy. He was the recipient of the best young researcher in the Arab Student Forum Creative among 61 researchers from 16 countries at Alexandria University, Egypt, in 2011. He has published more than 200 research papers in high-quality peer-reviewed journals and international conferences. He has also performed more than 600 reviews for high prestigious journals, including *IEEE Transactions on Power Systems, IEEE Transactions on Smart Grid, IEEE Transactions on Industrial Informatics, IEEE Transactions on Industrial Electronics, Energy Conversion and Management, Applied Energy,* and *International Journal of Electrical Power & Energy Systems.* He has participated in more than 15 industrial projects. His major research interests are power systems, power system dynamics, power system operation, control, dynamic state estimation, frequency control, smart grids, micro-grids, demand response, and load shedding.

Seifeddine Ben Elghali is an associate professor of electrical engineering at Aix-Marseille University, Marseille, France, since 2010. He received a BSc degree in Electrical Engineering in 2005 from ENIT, Tunis, Tunisia; an MSc degree in Automatic Control in 2006 from the University of Poitiers, Poitiers, France; and the PhD degree in Electrical Engineering in 2009 from the University of Brest, Brest, France. After receiving the PhD degree, he joined the French Naval Academy, Brest, France, as a teaching and research assistant. His current research interests include smart grids, virtual power plants, demand response, machine learning, renewable energy integration, energy management systems, and electric machines and drives.

MATLAB® is a registered trademark of The Math Works, Inc. For product information, please contact:

The Math Works, Inc.
3 Apple Hill Drive
Natick, MA 01760-2098
Tel: 508-647-7000
Fax: 508-647-7001
E-mail: HYPERLINK "mailto:info@mathworks.com" info@mathworks.com
Web: http://www.mathworks.com

Contributors

V. Arenas-Ramos
Departamento Ingeniería Electrónica y
de Computadores
Universidad de Córdoba
Córdoba, Spain

S. Bahramara
Department of Electrical Engineering
Sanandaj Branch, Islamic Azad
University
Sanandaj, Iran

Seifeddine Ben Elghali
Laboratory of Information & Systems
(LIS-UMR CNRS 7020)
Aix-Marseille University
Marseille, France

Nikos Bogonikolos
Blockchain2050 BV
Rotterdam, the Netherlands

Elutunji Buraimoh
Department of Electrical Power
Engineering
Durban University of Technology
Durban, South Africa

Innocent E. Davidson
Department of Electrical Power
Engineering
Durban University of Technology
Durban, South Africa

Z. Dehouche
Institute of Energy Futures Brunel
University London
Uxbridge, United Kingdom

Dah Diarra
RDIUP
Les Mureaux, France

R. Garner
Brunel University London
Uxbridge, United Kingdom

Jose Francisco Gomez Gonzalez
Departamento de Ingeniería Industrial
Universidad de La Laguna
La Laguna, Tenerife, Spain

Benjamín González-Díaz
Departamento de Ingeniería Industrial
Universidad de La Laguna
La Laguna, Tenerife, Spain

M. Gonzalez-Redondo
Departamento Ingeniería Electrónica y
de Computadores
Universidad de Córdoba
Córdoba, Spain

Ehsan Heydarian-Forushani
Department of Electrical and Computer
Engineering
Qom University of Technology
Qom, Iran

Dimosthenis Ioannidis
Information Technologies Institute
Centre for Research and Technology
Hellas
Thessaloniki, Greece

G. Jansen
Brunel University London
Uxbridge, United Kingdom

M. Lakshmi Swarupa
CVR College of Engineering
Hyderabad, India

Elpiniki Makri
Information Technologies Institute
Centre for Research and Technology
   Hellas
Thessaloniki, Greece

Juan Albino Méndez Pérez
Departamento de Ingeniería Informática
   y de Sistemas
Universidad de La Laguna
La Laguna, Tenerife, Spain

Entrit Metai
Blockchain2050 BV
Rotterdam, the Netherlands

I. M. Moreno-Garcia
Departamento Ingeniería Electrónica y
   de Computadores
Universidad de Córdoba
Córdoba, Spain

P. M. Naina
Indian Institute of Technology Madras
Chennai, India

Habib Nasser
RDIUP
Les Mureaux, France

Oleksandr Novykh
Departamento de Ingeniería Industrial
Universidad de La Laguna
La Laguna, Tenerife, Spain

V. Pallares-Lopez
Departamento Ingeniería Electrónica y
   de Computadores
Universidad de Córdoba
Córdoba, Spain

Meysam Qadrdan
Cardiff University
Cardiff, United Kingdom

R. Real-Calvo
Departamento Ingeniería Electrónica y
   de Computadores
Universidad de Córdoba
Córdoba, Spain

Saif S. Sami
Cardiff University
Cardiff, United Kingdom

I. Santiago
Departamento Ingeniería Electrónica y
   de Computadores
Universidad de Córdoba
Córdoba, Spain

K. Shanti Swarup
Indian Institute of Technology Madras
Chennai, India

K. Shashidhar Reddy
CVR College of Engineering
Hyderabad, India

P. Sheikhahmadi
Department of Electrical and Computer
   Engineering
University of Kurdistan
Sanandaj, Iran

Georgios Skaltsis
Information Technologies Institute
Centre for Research and Technology
   Hellas
Thessaloniki, Greece

G. Sree Lakshmi
CVR College of Engineering
Hyderabad, India

Contributors

Igor Sviridenko
Department of Energy Facilities of
 Ships and Marine Structures
Maritime Institute
State University of Sevastopol
Sevastopol, Russia

Christos Timplalexis
Information Technologies Institute
Centre for Research and Technology
 Hellas
Thessaloniki, Greece

Konstantinos Tsiomos
Blockchain2050 BV
Rotterdam, the Netherlands

Dimitrios Tzovaras
Information Technologies Institute
Centre for Research and Technology
 Hellas
Thessaloniki, Greece

Dmitry Vilenchik
Frish Technologies Ltd.
Kiryat Motzkin, Israel

Jianzhong Wu
Cardiff University
Cardiff, United Kingdom

Yue Zhou
Cardiff University
Cardiff, United Kingdom

Stylianos Zikos
Information Technologies Institute
Centre for Research and Technology
 Hellas
Thessaloniki, Greece

Introduction

This book discusses the virtual power plant concept and describes general approaches to analyze the need for and provision of additional flexibility based on virtual power plant in future power systems at both planning and operational time frames. Each chapter begins with the fundamental structure of the problem required for a rudimentary understanding of the methods described.

Chapter 1: Traditional passive customers are becoming "prosumers," proactive consumers with dispersed energy resources who actively manage their energy use, production and storage. As distributed energy resources (DERs) like distributed renewable energy storage, electric vehicles (EVs) and regulated loads become more prevalent, they have a profoundly disruptive and transformational impact on the central power system. The integration of DERs requires a paradigm shift to a decentralized power system with bidirectional power flow, which is widely accepted. They allow intelligent energy usage in a distributed environment by balancing demand and production in virtual power plants (VPPs). Another major advantage of VPPs is the integration of EV load management, which combines storage systems with regulated loads provided by the V2G service. To manage the VPP systems, a disturbed optimization method of its energy resources will be explored to optimize the management and scheduling. The development of a blockchain-based energy management platform for a VPP takes energy trading among users and network services provided to the grid into consideration. As part of the evaluation process for the proposed blockchain-based VPP energy management platform, a network of smart meters must be built as a proof-of-concept for connecting vehicles to the grid and grids to vehicles. The application of blockchain technology in the energy sector will be examined using a real-world example of peer-to-peer trading in electrical car charging. While a prosumer may use a rooftop solar plant to charge an EV, the prosumer's neighbor would use electricity from a faraway centralized power plant to charge his or her EV. If, on the other hand, the car is charged from a nearby solar power plant, the prosumer with solar will get a "buy-back rate" for the energy pumped into the grid.

Chapter 2: The smart grid enables suppliers and consumers to communicate systematically, making them more flexible and intelligent in their operating plans. A smart grid is an electrical network that monitors and manages the transportation of power from all generation sources to satisfy the variable electricity demands of end users using digital and other modern technology. Given the development of smart grids, demand-side participation, and microgrid's distributed energy resources integration into a single framework, such as a virtual power plant (VPP), it promises to reduce peak loads. Therefore, microgrids are critical parts of the future smart grid. This work discusses the present and envisaged prospective contributions of the VPP with the context of microgrids. This work presents a broad overview of the VPP of renewable energy with its attendant communication platform design and architecture. Thus, this work proposed a demand response control model and microgrid customers' electricity consumption model. Lastly, this work also discussed the electricity market interaction of the microgrid in demand response bidding.

DOI: 10.1201/9781003257202-1

Chapter 3: Exploration and analysis of the formation and optimal operation of virtual power plants (VPPs) that are providing services in a local flexibility market within an energy community, as well as in external electricity markets, is a topic of high value for smart grids. This chapter presents the design concept, characteristics and technical requirements of a system running at the aggregator level for formulating VPPs dynamically in an automated way and exploiting aggregated small-scale distributed energy resources' flexibility in smart energy communities. The created flexibility bundles are utilized to better exploit prosumers' flexibility and generate profit for prosumers through trading in electricity markets (e.g. local flexibility market, ancillary market). Moreover, a methodology for dynamic VPP formation is presented, which is driven by the grid network state that is defined according to the traffic light concept. A key point is that the input data that are considered by the dynamic VPP formation method depend on the current grid state. A proposed heuristic optimization algorithm was tested via simulation experiments in a use case, demonstrating the ability to find good solutions in a very short time.

Chapter 4: Renewable distributed energy resources (DER) are becoming more popular as governing bodies introduce ambitious climate policy objectives to curb emission production. A community-driven virtual power plant (VPP) concept can be used to increase the visibility of DER and provide grid support services and flexibility, as well as promote the continual increase in renewable energy usage and additional revenue to participants. In this model, a prosumer energy community is constructed consisting of surplus photovoltaic solar generation, community loads, and a hybrid battery and hydrogen fuel cell (HFC) energy storage system (ESS). The devised optimization process concluded that hybridizing a 6-kW fuel cell and 20-kW electrolyzer with a 50-kWh lithium-ion fuel cell provides optimal synergy between the technologies. This strategy makes use of the strength of each technology by allowing the battery to provide short-duration storage, while the HFC can provide long-duration storage and improve the overall resilience and economic performance of the ESS. The hybrid storage is sized to reduce the levelized cost of electricity (LCOE) by finding the optimal battery and HFC capacity ratio, with a value of 17.6 €cent/kWh. Modeling results show a considerable reduction in the carbon impact of the overall system through maximizing the utilization of clean energy resources ranging from 36% to 58% emission intensity for the different included buildings, as well as the decrease in the LCOE over the 20-year system life span. Further research of the system's economic performance and potential business models in combination with this work could prove the commercial viability of the concept of energy communities.

Chapter 5: In the context of future smart grids, the distributed and small-scale energy resources will play a more remarkable role in comparison with their present situation. One of the main technical and commercial challenges for the integration of distributed energy resources (DERs) into the grid is their invisibility from system operator viewpoint. Virtual power plant (VPP) concept seems an appropriate solution to tackle such a problem. In fact, a VPP is composed of small DERs (traditional and renewables), controllable loads and energy storage devices and can coordinate them in order to provide several local and system services through participation in different electricity markets with the aim of maximizing its profits. This chapter

Introduction

presents appropriate models for optimal energy management within a VPP that could be a technical VPP (TVPP), a commercial VPP (CVPP) or an energy community. The effectiveness of the presented models has been validated through several numerical analyses.

Chapter 6: The transition to a low carbon power system is facing unprecedented challenges, with the high penetration of converter connected and distributed renewable generation and thus rapidly increasing demand due to electrification of heat and transport. In this chapter, a smart energy management paradigm, called a virtual energy storage system (VESS), is presented to address these challenges and support the cost-effective operation of future power systems. The VESS concept is defined first, followed by discussions about the related enabling technologies, control schemes, possible applications and potential benefits. Then an overview of components of a VESS, including flexible demand and conventional energy storage systems (ESS), is presented. Two different applications of the VESS are demonstrated, i.e. frequency response services to the power system operator and voltage support to distribution network operators.

Chapter 7: With energy evolving before our eyes from the era of fossil fuels and their effects on the earth to renewable energy sources and energy storage and management, it would be wise to seek out an innovative way of reducing the hassle of the management aspect by introducing blockchain-enabled solutions and smart contracts functionalities. It is important to lessen third-party interferences and implement much smoother executions of contract agreements where all parties are abreast of all details, changes and conclusions as they occur. In the energy sector, accuracy, trust, security and saving costs are paramount, and these are the major advantages of smart contracts linked to the blockchain. Contemporary energy management systems, which usually involve the generation of orders, trade compliance, managing orders, price delivery, exchange execution and settlement accounting, are all time-consuming. The lack of flexibility allows for too many complications tying in several intermediaries. For grid operators, blockchain technology, smart contracts and decentralized software can be utilized to create a seamless, secure and efficiently distributed energy system promising to solve the majority of these highlighted pitfalls.

Chapter 8: The pollution emission of fossil-fuel-based power plants has a major effect on global warming. Although producing electrical energy through renewable energy sources (RESs) is introduced as an appropriate solution, integrating these resources into power systems faces several challenges. To manage these challenges, different innovations are proposed. One of these innovations, which is defined under the business model category, is using aggregators for the optimal operation of distributed energy resources (DERs). Many DERs can be integrated under the aggregator's management to participate as a virtual power plant (VPP) in the wholesale markets. The VPP aggregator (VPPA) can sell energy to the day-ahead and real-time energy markets and it can provide different services for the grid operator through the ancillary service markets. Therefore, appropriate decision-making frameworks are required for the VPPA to model its bidding strategies in the wholesale energy and ancillary service markets on one hand and to model the VPPA's behavior within its resources on the other hand. For this purpose, after describing different frameworks

4 Virtual Power Plant Solution for Future Smart Energy Communities

for the VPPA's decisions in the wholesale markets in this chapter, a review of the previous studies is done considering the decision-making frameworks, mathematical formulations, uncertainty models and solution approaches.

Chapter 9: The book chapter discusses a decentralized energy management algorithm to control the participants, such as renewables, energy storage, distributed generations, and controllable loads in VPP (virtual power plant). The VPP enables the participation of small distributed energy resources (DERs) in electricity markets by improving their profits. In this work, the VPP operation problem is formulated as the maximization of economic benefit subjected to power balance constraints, network constraints and local constraints of DERs. There are two ways for transmitting information within VPP: the centralized approach and the decentralized approach. Since the VPP has no geographical limitations, a centralized algorithm needs a powerful central controller to collect and process the data and also requires a high bandwidth communication network. Hence, a robust distributed algorithm is necessary to get an accurate, efficient operation. A communication-based algorithm is proposed to reach a consensus among the operating variables, which is robust to communication delay, noise and communication failure. The faster convergence enables the use of the algorithm in real-time markets. The results for a 15-node VPP system were presented in detail.

Chapter 10: The grid-edge concept promotes the proximity of energy to consumers to the detriment of energy produced in conventional power plants. This new model helps to reduce the losses associated with distribution lines by promoting microgeneration with the participation of consumers. Coordinated control and monitoring are necessary for microgrids to behave as a single generation system, similar to conventional energy sources. In this context, maintaining a unique and high-quality time reference frame is essential. In this connection, this chapter systematically addresses the study of two synchronism technologies based on Precision Time Protocol and Time-Sensitive Network protocol. Through the exhaustive analysis, the conclusion is drawn that a synchronization scheme based on both protocols can be applied to new infrastructures dedicated to the monitoring of microgeneration power grids and could be applied to larger areas with the participation of μPMU. The proposed infrastructure works with standard Ethernet networks so that measurement equipment can coexist with other data transmission systems in various production and consumption scenarios. In practice, this means that standard Ethernet functionality is enhanced to guarantee message latency for both critical and noncritical traffic. The aim is to ensure real-time control outside the local area of operations.

Chapter 11: A clean virtual power plant (VPP) aggregates small distributed renewable energy resources (DRES) to act together as a single power plant with the possibility to store the surplus or release the electricity according to the system's needs in an effective way. However, integrating renewable energy source (RES) especially wind and solar in energy grids introduces variabilities and uncertainty issues that need an appropriate demand response management and proposer planning and coordination. The two most challenging parts for their penetration are the complementarity between the VPP components and the flexibility capacity that should be provided through the controllable loads, storage systems and energy sources. This chapter presents a review of the main research topics revolving around

Introduction

the optimization of portfolio and proposes an interoperable application programming interface that estimates key metrics for VPPs. Particularly, we present a complementarity index that calculates the level of synergy between VPP components based on the correlation between different energy systems and a flexibility factor that computes the capability of a VPP to maintain a balance between generation and load during uncertainty. Furthermore, we discuss the robustness of these indexes by simulating some scenarios with different levels of variability and uncertainty and we provide insights into future research directions.

Chapter 12: The Gorona del Viento hybrid power plant project, implemented in 2014 on the island of El Hierro, is a successful example of the use of renewable energy sources to supply electricity to a large number of consumers. The operating experience of this hybrid power plant is quite important for the further development of such projects. It should be especially noted that the parameters of all elements of this hybrid power plant are publicly available on the Red Eléctrica de España website in real time, which allows for a detailed analysis of operating experience, as well as to check the effectiveness of using various technologies for active control of an isolated power system in order to optimize its parameters and improve its efficiency. The increase in renewable and distributed energy resources complicates efficient grid management. For these purposes, the technology of virtual power plants is best suited in combination with sufficiently effective systems for short-term power forecasting and active consumer management. This work is devoted to the presentation of the experience of building a virtual power plant to work in conjunction with a hybrid power plant "Gorona del Viento," located on the island of El Hierro. For these purposes, the chapter presents a detailed analysis of the operating experience of the hybrid power plant "Gorona del Viento" over the past three years. It also describes the developed software package for predicting the load in the electrical network using artificial intelligence technologies, neural networks, deep machine learning and simulation (digital twins).

The Editors

Ehsan Heydarian-Forushani, Hassan Haes Alhelou, and Seifeddine Benelghali

1 Practical Implementation of VPP in the Real World Based on Emerging Technologies

M. Lakshmi Swarupa, G. Sree Lakshmi, and K. Shashidhar Reddy
CVR College of Engineering, Ibrahimpatnam, Hyderabad, India

CONTENTS

1.1 Introduction to VPPs 8
1.2 Advantages of VPPs 9
1.3 Disadvantages of VPPs..10
1.4 Elements of VPPs 10
1.5 Ideal Virtual Power Plant 11
1.6 Types of VPPs..14
    1.6.1 Differences of VPPs Due to Location on the Grid............................14
1.7 Case Study of VPP in Real World 15
    1.7.1 Information Communication Technology (ICT) 16
    1.7.2 One Solution for All 18
1.8 Generic Benefits From VPPs 18
    1.8.1 Advantages for TSOs..19
    1.8.2 Advantages for DSOs 19
1.9 Emerging Technologies in VPP...20
    1.9.1 Integrated Distributed Grid Control to Facilitate Peer–Peer Energy Trading 20
    1.9.2 Design of a Peer–Peer Energy Trading Platform 20
    1.9.3 Control of an LV Microgrid with Peer–Peer Energy Trading 22
    1.9.4 Energy Management of Different DERs by Using Optimization Algorithms...24
1.10 Blockchain Technology 25
    1.10.1 Types of Blockchain Techniques 27
    1.10.2 Microgrid Applications of Blockchain Technology 28
    1.10.3 Dynamic of Energy Consumption, Production, and Exchange..........29

DOI: 10.1201/9781003257202-2

7

8   Virtual Power Plant Solution for Future Smart Energy Communities

1.11 Case Studies...31
    1.11.1 IEEE 14 Bus Standard System...32
    1.11.2 Existing Policies and Regulations Indirectly Supporting the
           Adoption of VPP in India 36
References...38

1.1 INTRODUCTION TO VPPs

Traditional passive customers are becoming proactive consumers with dispersed energy resources who actively manage their energy use, production, and storage. As distributed energy resources (DERs) like distributed renewable energy storage, electric vehicles (EVs), and regulated loads become more prevalent, they have a profoundly disruptive and transformational impact on the central power system. The integration of DERs requires a paradigm shift to a decentralized power system with bidirectional power flow, which is widely accepted. They allow intelligent energy usage in a distributed environment by balancing demand and production in virtual power plants (VPPs) [1]. Another major advantage of VPPs is the integration of EV load management, which combines storage systems with regulated loads provided by the vehicle-to-grid (V2G) service. To manage the VPP systems, a disturbed optimization method of its energy resources will be explored to optimize the management and scheduling [2].

The development of a blockchain-based energy management platform for a VPP takes energy trading among users and network services provided to the grid into consideration. As part of the evaluation process for the proposed blockchain-based VPP energy management platform, a network of smart metres must be built as a proof-of-concept for connecting vehicles to the grid and grids to vehicles. The application of blockchain technology in the energy sector will be examined using a real-world example of peer-to-peer (P2P) trading in electrical car charging. While a prosumer may use a rooftop solar plant to charge an EV, the prosumer's neighbour would use electricity from a faraway centralized power plant to charge his or her EV. If, on the other hand, the car is charged from a nearby solar power plant, the prosumer with solar will get a "buy-back rate" for the energy pumped into the grid [3]. Depending on the particular market environment, VPPs can accomplish a whole range of tasks. But the VPP not only helps stabilize the power grids it also creates the preconditions for integrating renewable energies into the markets. The dependable supply of solar energy produced by a VPP not only helps stabilize the electricity grid (meaning less blackouts in peak demand times), but it's also a more sustainable energy solution than a conventional power plant. A VPP reduces reliance on coal power.

As DERs grow in number, new technologies, energy strategies, and laws are needed to cater to the technical and economic issues that emerge because of this growth. To handle the integration of distributed generation (DG), the idea of a VPP arose. A combined power plant, which is another term for a VPP, consists of many small power plants scattered throughout the system, as well as load controllers and energy storage devices. Because of this, as shown in Figure 1.1, the VPP makes use of both traditional fuels like coal and nuclear power while also using renewable

Practical Implementation of VPP

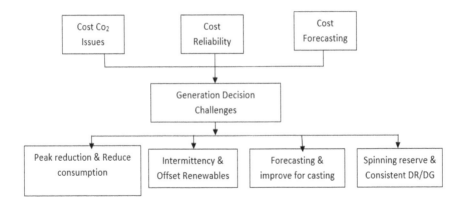

FIGURE 1.1 Renewable intermittency.

energy sources (RESs) like photovoltaic (PV) power plants, wind turbines, and modest hydro. All these facilities are controlled by a single cloud-based remote-control centre using the VPP software.

Dispersed production locations and storage facilities may be linked, monitored, and coordinated more efficiently using the VPP software. The cloud-based control centre makes use of information and communication technologies (ICTs) and the Internet of Things (IoT) to manage and show the DG units as a single large central power plant. VPPs are an aggregation of consumers (residential, commercial, or industrial) that are subject to a single kind of pricing, the demand objective of a DERs programme. While the idea seems quite like what we have now, the major difference is that a VPP is specified at a finer level than the entire programme. Utilities are no longer required to put all customers with a certain programme under one roof. The idea of VPP enables utilities to combine these programmes based on their type and placement in the distribution topology, or any other agreed-upon aggregation [4].

1.2 ADVANTAGES OF VPPs

- Generators that will not be affected by a weather change are cranked up, while those that will be affected are turned down or shut off based on predictive predictions from VPPs. As an example, if a cloud cover prediction is expected over a solar farm, the management system would increase the power output from bioenergy plants to keep total output at a constant level and offset intermittent concerns.
- To produce hydrogen, use excess renewable energy. Then, store the hydrogen until it's needed to meet the gas network heat demand.
- During high demands, increase the capacity of the independent generator.
- Flexible power consumers (such as an industrial pump) may be ramped up or down depending on market prices, allowing for energy consumption when it is cheap and demand is low.
- Grid frequency management and general grid balancing services are provided by flexible power sources, such as bioenergy and hydropower plants.

Existing VPPs are building the foundation for a decentralized energy system, which is promising. This new energy Internet provides numerous advantages to grid operators, power producers, and power consumers in general. It is obvious that intermittency problems may be handled by employing advanced algorithms and linking together independent generators [5].

- DERs are used to offer service to the distribution and transmission networks in the north.
- Losses will be minimized.
- A management system will be used to identify failures.
- System voltage and frequency control are auxiliary.
- Proactive will gain financially.
- Network operation has a higher visibility of DER units for consideration.
- Economic integration of renewable energy on a large scale is made while preserving system security.
- Energy marketplaces are open to small-scale players.
- There is an increase in the overall efficiency of electric power.
- Possibilities for new company ventures are achieved.

1.3 DISADVANTAGES OF VPPs

- Cyberattacks pose a threat to them.
- VPP's second issue is that, unless a voltage management system is in place, large amounts of DERs may influence local voltage.
- VPPs have regulatory requirements, and incentives are required to assist VPP operators in bringing their advantages to the system.

With the introduction of VPP grids and their potential to produce massive amounts of data on consumers, including their energy consumption and trends, as well as other personally identifiable information (P11), security has never been more important. This is another area where security should be prioritized. The terms "cyber terrorism" and "cyber security" are interchangeable. It has pros and cons because of advancements in information communication technology (ICT) and the high cost of using industrial control systems (ICS) in the energy industry. The energy industry performs a fantastic job of controlling the risks that its operations face.

Table 1.1 shows the similarities between conventional and smart grids, while Table 1.2 shows the comparisons between VPP and microgrids.

1.4 ELEMENTS OF VPPs

Figure 1.2 shows a VPP system; the VPP connects many different power sources to provide a stable power supply. When different types of distributable, non-distributable, controllable, and load-flexible (DG) systems are used together, they form a centrally managed group. These systems are transferable, non-distributable, controllable, or load-bearing DG systems (Energy Storage System) [6]. Utilizing existing grid networks to create customer supply and demand services, virtual power stations, as

TABLE 1.1
Comparison between Traditional Grid and Smart Grid

Component	Traditional Grid	Smart Grid
Energy generation	Nuclear, hydroelectric, thermal, and fossil-based fuel centralized energy generation	Wind, solar, biomass, and geothermal energy are examples of dispersed energy production resources
Technology	Analogue – Electrochemical, solid-state devices, bidirectional, weak, and unreliable	Microprocessor-based, robust, and dependable, broad ICT usage bidirectional, global
Communication	Unidirectional, bidirectional	Bidirectional, global
Business model protection data	Static is restricted for consumers, and manual restoration is limited	Processors that are dynamic, adaptive, and self-healing are plentiful

TABLE 1.2
Comparison between a Virtual Power Plant and Microgrid

Sl. No.	Virtual Power Plant	Microgrid
1.	The grid is interconnected with VPPs	Microgrids are often off-grid, and they are intended to be islanded in an on-grid environment so that they may continue to function independently if the grid goes down
2.	Assets linked to any section of the grid may be used to create VPPs	Microgrids are typically limited to a certain area, such as a neighbourhood or an island
3.	VPPs are controlled by aggregation software, which provides capabilities like those found in conventional power plants	For islanding, on-site power flow, and power quality control, microgrids depend on extra hardware-based inverters and switches
4.	VPPs are often used in wholesale markets and do not need special regulation	Microgrids, on the other hand, are primarily concerned with providing electricity to end users

shown in Figure 1.3, are called "Internet of Energy." A wide range of software-based technologies is used by VPPs to maximize value for both the end consumer and the distribution service provider. When it comes to responding to changing customer load conditions, VPPs are responsive, adaptable, and offer value in real time.

1.5 IDEAL VIRTUAL POWER PLANT

The ideal VPP is made up of two major components:

A. Technology generation

The DG standard is helpful for describing the capabilities of different technologies that are usually classified as DG. DER is being explored for inclusion in VPP [7].

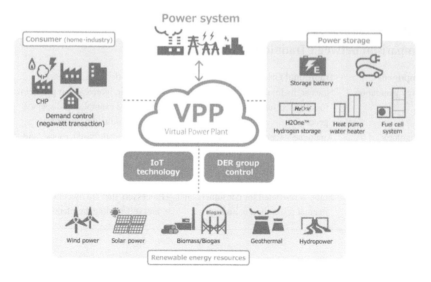

FIGURE 1.2 Elements of VPP.

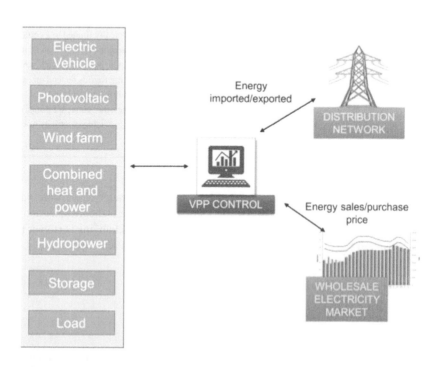

FIGURE 1.3 Diagram of VPP and its interaction with electricity markets and networks.

Practical Implementation of VPP

- CHP (combined heat and power)
- Biomass and biogas
- Power plants with a smaller capacity (gas turbines, diseases, etc.)
- Plants that produce water from a little source
- Power production from the wind
- Production of solar energy
- Consumption that is adaptable (loads that can be sent)

In this regard, all DGs may be divided into two groups which are defined as follows:

1. Domestic distributed generator (DDG): It is a small DG unit that provides power to individual consumers in residential, commercial, and industrial settings. A DDG owner's excess power output may be fed into the grid, and its shortfall can be compensated by the system.
2. Public distributed generator (PDG): The primary purpose of a DG unit that does not belong to a single customer is to supply energy to the grid. In general, energy storage may be installed in DDG and PDG. It is an energy storage system connected to a loaded generator and a low-voltage distribution network, called a DDG. A generator and energy storage facility like PDG can only imply one thing: When it comes to power and, most likely, heating, DDG owners want to keep their costs down while still getting reliable service. Most essential business regulations are unknown to them. While PDG owners want to sell their electricity to the grid, it's not their primary aim to do so. Compared to PDGs, DDG generation capabilities are generally weak. For this reason, the crane with DDG can participate as an individual partner in the energy market, but not with PDG. Wind and solar systems without energy storage are stochastic PDGs (SPDGs) or DDGs. Fuel cells (FCs) and microturbines, on the other hand, are dispatchable, which means they can quickly switch between operating modes. For the former, there are dispatchable PDGs (DPDGs) and SPDGs, whereas for the latter, there are dispatchable DDGs (DDDGs) and stochastic DDGs (SDDGs).

B. **Technology for energy storage**

It's now widely accepted that energy storage devices may help balance out variations in power demand with a given level of power supply. Although wind turbines and PV technologies aren't intended to be used as primary sources of energy, they may be supplemented with them or used as temporary energy storage. The incorporation of Energy Storage System in VPP is currently being investigated.

- DER units' visibility in energy markets
- DER units' participation in energy markets
- Maximizing the value of DER unit's involvement in the energy markets

Comparing portfolio variety and capacity to stand-alone DER units enables tiny units to enter markets and minimizes the danger of imbalance. Active distribution networks (ADNs) must address network operational characteristics for long-term stability when commercial VPPs (CVPPs) perform commercial aggregation. The combined DER units may be deployed across several distribution and transmission grids without regard to location. Thus, a single region of a distribution network may include multiple CVPP aggregating DER units. In general, CVPP duties should comprise the following:

- Characteristics of DERs are maintained and submitted
- Forecasts for production and consumption
- ODM (outage demand management)
- DER proposals for construction
- Bids are submitted to the market
- Optimization and generation scheduling daily
- DERs are selling energy providers to the market

1.6 TYPES OF VPPs

The two types of VPPs are mentioned as shown in Figure 1.4: (1) Technical VPP (TECHNICAL VIRTUAL POWER PLANT(TVPP)) and (2) CVPP.

1.6.1 DIFFERENCES OF VPPs DUE TO LOCATION ON THE GRID

There are different views on VPPs with respect to generation, transmission, and distribution level. There is differentiation among so-called local VPPs (LVPPs), regional VPPs (RVPPs), and large-scale VPPs (LSVPPs). Figure 1.5 illustrates the operational architecture of the Bakari Project depending on the location of VPP. The architecture is a general approach and is abstracted from country-specific features.

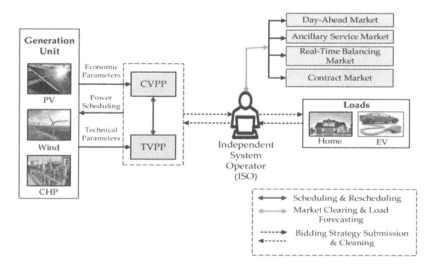

FIGURE 1.4 Integration of VPPs in deregulated power market.

Practical Implementation of VPP

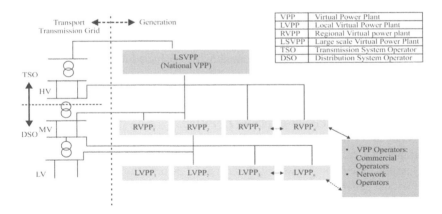

FIGURE 1.5 Different types of LVPP, RVPP, and LSVPP.

The different types of LVPP, RVPP, and LSVPP can be seen as consecutives.

- The operation of an LVPP is performed by a local energy company or some other local energy market participants. The intended purpose of an LVPP is optimal usage within an electricity market (for example, EPEX/EEX SPOT) as well as different kinds of network support (congestion management, voltage support) or system services like balancing services.
- On the next level, RVPPs consist of an aggregation of LVPPs and are operated by the distribution system operator (DSO) or other commercial parties. They're in charge of managing the regional distribution network's electricity flows.
- A VPP controlled at the transmission level and therefore under the control of the transmission system operator (TSO) is produced by merging these RVPPs in an LSVPP.

1.7 CASE STUDY OF VPP IN REAL WORLD

The number of RESs and electrical energy storage (EES) has steadily grown in contemporary electrical power networks [8]. As a result, new methods to managing them, like microgrids and VPPs, have emerged. In general, VPPs are self-contained devices with efficient power flow management systems. VPPs consist of several components that are all linked to the distribution network. Generators, loads, and energy storage systems are the components mentioned. Coordinating the efforts of the whole VPP is a tough and time-consuming task.

- Hydraulic pumped energy storage (HPES)
- Compressed air energy storage (CAES)
- Flywheel energy storage (FWES)
- Superconducting magnetic energy storage (SMES)
- Battery energy storage system (BESS)
- Supercapacitor energy storage (SCES)
- Hydrogen along with Fuel Cell

1.7.1 Information Communication Technology (ICT)

Communication technology and infrastructure are key requirements for VPP. Media technologies may be regarded as communications including energy management systems (EMSs), supervisory control and data acquisition (SCADA), and distribution dispatching centre (DCC).

A. Technical virtual power plant (TVPP)

Technical Virtual Power Plant (TVPP) was created with DER from the same geographical area. With TVPP, one can see how the local network affects the entire DER profile in real time as well as the cost and functionality of the portfolio. TVPP services and activities include DSO local system management, as well as TSO system balancing and ancillary services (TSOs). The operator of a TVPP, usually the DSO, needs comprehensive information about the local network. The TVPP makes it possible to achieve the following:

- DER units are visible to system operators.
- DER unit's contribution to system management.
- Optimal use of DER unit's capability to offer ancillary services while considering local network limitations.

By diversifying portfolios and capacity, small DER units may provide additional services while reducing unavailability concerns. The technical control capabilities of distributed generators are examined in depth, as well as the auxiliary services that may be provided consequently. The technical potential is examined using a novel evaluation method that analyses each Grid-coupling converter individually, as well as its unique capabilities. A huge technical potential has been discovered. DSOs that use the TVPP model are also known as ADN operators. An ADN operator may utilize DER units' auxiliary services to improve the performance of their network. An ADN operator, on the other hand, may provide auxiliary services to other system operators. ADNs may be organized in a hierarchical or parallel manner based on different voltage levels or network regions when the TVPP concept is used. The Active Network Deployment Register has several examples of ADNs. TVPP is responsible for a variety of functions, including:

- Continual condition monitoring – Recovery of previous equipment loadings
- Asset management that is backed up by statistics
- Self-description of system components and self-identification
- Fault location – Linked with outage management automatically
- Maintenance was made easier
- Statistical analysis and portfolio optimization of projects

B. Commercial virtual power plant (CVPP)

CVPPs contain a profile and total output indicating the cost and operational characteristics of the DER portfolio. The CVPP profile does not take the distribution network's impact into consideration. CVPP's services and responsibilities include

TABLE 1.3
Current Research Trends Concern Virtual Power Plants (VPPs) and Power Quality (PQ)

Current Research Trend Concerning VPP

- Optimal active and reactive power scheduling network voltage control by renewable energy sources integrated power flow control and analysis
- Electric energy storage sizing, localization, and management in BPP
- Playing a role in energy market energy management in a VPP – Real case study analysis

Current Research Trends Concerning PQ

- Development of PQ measurement devices and systems
- Data mining of PQ measurements
- Global power quality indices
- Detection and classification of voltage events
- Vehicle-to-grid (V2G) impact on PQ

---

trading in the wholesale energy market, balancing trading portfolios, and providing services to the [transport] system operator (through bids and offers). Anyone who has market access, such as an energy provider, can run an aggregator or a balancing responsible party (BRP). This is a difficult job. The functioning of VPPs may be studied in a variety of ways. Table 1.3 presents the current research directions concerning VPPs and power quality (PQ).

VPPs can also be used as vehicles for integrating DERs into the current power system at a low cost. The case studies show how the ideas are applied to a test system. The system's performance is measured in terms of energy efficiency, PQ, and security. The IEC 61850 standard has been extended to improve the interaction between the VPP controller and the DER. In a real-world scenario, the VPP communication and control architecture is implemented to provide a suitable framework for harmonizing the operations of different units of VPP. The decisions and the generation of profit, although complying with the required PQ levels and physical network constraints, are important elements of VPP strategies. In mathematical modeling, a VPP consists of base load thermal power plant on an islanded grid, which also includes 200 MW of wind energy, 100 MW of solar energy and + or − 250 MW of pumped storage are considered. This chapter's goal is to demonstrate the impact of variable power generation on PQ by outlining several approaches to managing storage plants.

The issue of multi-energy VPP coordinative operation is addressed. The bi-objective dispatch method was developed for optimizing multi-energy VPP performance in terms of economic cost and PQ. The Hong Feng Eco-town in Southwestern China was the subject of a genuine case study. The compromised approach optimizes VPP management with priority needs. The fuzzy multiple objective optimization problems were used to construct the VPP's operation optimization model. This optimization issue includes customer and supplier satisfaction, system stability, PQ, and costs with operating constraints. The technique that was suggested is used.

Increased renewable energy (RE) and DER penetration leads to various technical and economic issues in the energy system. Varying PQ due to fluctuating RE generation, grid congestion, non-predictability, and no market visibility are some of the issues arising with increasing RE and DER. To solve these problems, a VPP may be used. Distributed producing units, regulated loads, and storage devices work together to create a single VPP. Various research institutes define and classify the technology in different ways. VPP may be classified as technical and CVPP based on the end-use application, or local, regional, and large scale based on the area catered. VPP can basically be considered "software-as-a-solution," the architecture and cost of which depend on the use case applicable. The cost of a medium-sized VPP having aggregated control functionality can be expected in the range of $20,000–$300,000. The high cost of VPP may be attributed to the need for infrastructure (communication hardware) and license model for selling the VPP software. This cost may differ with varying licenses and varying number and type of consumers. However, since the technology is at an initial phase, giving an exact cost of the solution may be difficult.

1.7.2 ONE SOLUTION FOR ALL

There are numerous benefits associated with a VPP. For example, owing to metering installations on site, VPP may assist in improving DERs' visibility to all energy players and enable remote control capabilities of DERs owing to a robust communication network, which is presently lacking in the power system.

- For load dispatch centres: VPP can help in improving scheduling of RE plants, automated network monitoring, and accurate accounting of electricity transmitted. It can also provide ancillary services from aggregated DERs.
- For transmission companies: VPP can help in providing relief in the transmission congestion and transmission upgrade deferral.
- For distribution companies: VPP can help in providing accurate forecasting of net load and thus efficient planning of power purchase. Aggregated and visible DER in a VPP can also help distribution companies in reducing the quantum of power purchased.
- For customers having DER: VPP can provide market visibility to the customers and can help in improving their supply capacity to grid. For off-grid customers, aggregation of DER in a VPP can help them in performing P2P transactions in future.

1.8 GENERIC BENEFITS FROM VPPs

Several broad conclusions may be drawn about the advantages of VPP systems. As a result, it's critical to understand the key technological features that come with a VPP:

- Owing to on-site metering systems, DERs are visible to all energy participants.
- Remote control capabilities of DERs due to strong communication network, which grants access to DERs.

Practical Implementation of VPP

These two crucial technical requirements, DER visibility and controllability, may provide many benefits. Following is a discussion of the possible enabling advantages of VPP for a variety of electrical system stakeholders, including network operators (TSO and DSO), DERs operators, market aggregators, and regulators.

1.8.1 Advantages for TSOs

TSOs are responsible for ensuring supply security and good quality throughout the whole system. This necessitates accounting for the huge phase-out of conventional power. Large VPPs of decentralized DERs may assist in filling the gap by operating entirely at the transmission level, providing frequency responsiveness, voltage support, and black start capability.

1.8.2 Advantages for DSOs

In contrast to the present scenario, as the number of DERs linked to the distribution grid grows, a fundamental shift occurs: The responsibility for system stability is reversed. Because the bulk of installations of DERs is on the distribution level, the DSO's activities are critical for ensuring security in a renewable-driven electricity system. Controlling DERs remotely may help a DSO run more efficiently without adding any new capacity to the grid. This helps maintain a secure system. Operators of DERs will benefit from this: Aside from the benefits of VPPs in terms of system security, DERs may benefit from a variety of previously mentioned economic possibilities. High transaction costs for selling only one unit, as well as market limitations like a minimum offer limit or overly lengthy product durations, limit DER owners' ability to make larger profits. Market aggregators will benefit from this (German market): As a bridge between a VPP operator and Germany's electricity market, aggregators benefit from rising volumes of energy market commitments and the transaction fees that go along with them. We expect that with VPPs, there will be additional trade and procurement processes on top of the current direct marketing of larger DERs. As a result of internal balancing in a VPP pool owing to its own flexibility or increased short-term market liquidity, or an increasing cost position with higher penetration of wind and solar energy without countermeasures such as the introduction of VPPs, VPPs facilitate the reduction of costs for procuring balancing energy.

Regulators will benefit from this: Finally, three major advantages may be expressed in terms of the "energy triangle" for regulators: Economic profitability, supply security, and sustainability. As a result of lower grid expansion costs and reduced system flexibility expenses, end-user electricity prices have decreased, resulting in lower expenses for congestion management activities, lower expenses for reserves and short-term electricity products due to increased market offers, improved forecasts, and internal balancing. Second, it gives the regulator more control over the power system's flexibility, which improves the system's overall stability. In the battle against climate change, the VPP's controlled and renewable DERs contribute to the struggle, fulfilling the energy triangle's objective [9].

1.9 EMERGING TECHNOLOGIES IN VPP

- Forecasts by energy-generating technology component
- Technology for storing energy
- Wind-based energy generation uses ICT to generate energy
- Plants that produce hydrogen
- Production of solar energy
- Combined heat and power(CHP)
- Information communication technology (ICT)
- Energy management systems (EMS)
- Supervisory control and data acquisition (SCADA)
- Distribution management system (DMS)
- Battery energy storage system (BESS)
- Supercapacitor energy storage (SCES)
- Super conductor magnetic energy storage (SMES)
- Hydraulic pumped energy storage (HPES)
- Flywheel energy storage (FWES), other, in accordance with the kind of end user (industrial or commercial), as well as regional and national market sizing and examination of the top VPP providers

1.9.1 INTEGRATED DISTRIBUTED GRID CONTROL TO FACILITATE PEER–PEER ENERGY TRADING

Conventional energy consumers are turning into proactive as more DERs are linked. The power generated by DERs is often inconsistent and unexpected. Excess energy generated by consumers can be reduced, stored in energy storage devices, exported back to the grid, or sold to other energy consumers. P2P energy trading refers to direct consumer-to-consumer energy trading based on the concept of a "P2P economy" (also known as a sharing economy) and often local authority distribution network.

A peer is a single group or group of domestic energy customers such as generators and consumers in the context of P2P energy trading. Peers deal directly with one another to buy and sell energy without the involvement of conventional energy suppliers. P2P energy trading is often made possible via online services based on ICT.

Traditional energy trade is mostly one-way. Electricity is often delivered over vast distances from large-scale producers to users, whereas currency flows in the other direction. P2P energy trading, on the other hand, promotes multidirectional trade within a small geographic region. Using the "P2P economy" concept, energy trading experiments have already been conducted across the globe, as shown in Figure 1.6.

1.9.2 DESIGN OF A PEER–PEER ENERGY TRADING PLATFORM

The development of DERs and variable loads, as well as grid digitalization, is shaping the future of distribution networks (smart grids). A passive unidirectional circuit servicing consumer loads is not necessary, when DERs are integrated into the distribution network. It becomes a dynamic system capable of transferring power in both directions. Integration of DERs into distribution networks offers many benefits,

Practical Implementation of VPP

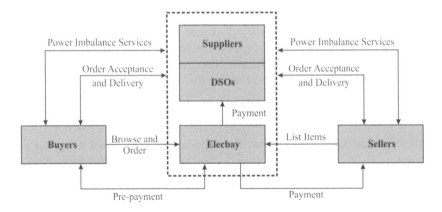

FIGURE 1.6 Peer-to-peer trading in microgrid.

including decreased greenhouse gas (GHG) emissions, loss reduction, and grid dependence reduction. However, as the number of DERs grows, current distribution networks face a variety of operational and market problems, including voltage limit breaches, line congestion, DER visibility issues, and intermittent energy imbalances. These technological problems are typically regional or local in nature. It is difficult to create a centralized market that serves many places to address such local issues due to a variety of obstacles. Local energy trading may be a feasible alternative for dealing with local issues that arise because of the integration of DERs into distribution networks [10].

Local energy trading, as shown in Figure 1.7, is a novel idea that is gaining traction in the field of distribution networks. The main idea of energy trading is to provide a market platform for participants who are capable of active market

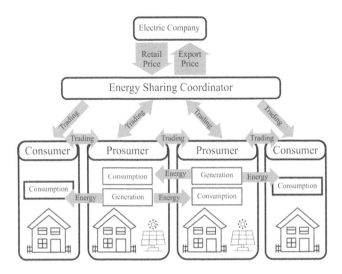

FIGURE 1.7 Peer-to-peer energy trading.

participation, where they may manage their needs or generation adaptively to meet electricity production and distribution capacities at any time and engage in the market as a seller or buyer. This market allows sellers and buyers in the distribution network to trade energy on a P2P basis. This aim can be performed by designing a responsive platform, which is a structure that enables market players to control their energy expenditures by changing their electricity use in response to electricity price. Players with renewable energy and electrical energy storage (EES) can respond to the market by managing their generation and charging/discharging EES and act as seller/buyer in the market. The designed platform should have three main features: Scalability, addressing network constraints, and low communication and computation overheads.

Scalability refers to the capacity of the planned platform to deal with and function in the face of a significant rise in the number of participants. The number of market participants will grow to hundreds of thousands as more DERs are incorporated into the electricity grid. As a result, the planned market should be able to handle many participants. Another significant obstacle that emerges when the number of market participants grows is the system's overhead. Communication and processing costs will be placed on the system with a high number of participants as part of these overheads. Owing to the growing processing load and communication infrastructure caused by an ever-increasing number of participants, the system will eventually reach a breaking point.

A market that overlooks grid constraints is inherently untenable as it fails to take both economic and technical considerations into account. Consequently, network constraints should be the next key consideration in the design of the local energy trading market. With this background in mind, this thesis offers two different market structures for the selling of P2P energy: A community-based market and a bilateral trading market. Auction-based technique, distributed optimization, and decentralized market clearing are some of the market's clearing methods that have been developed. In every individual transaction in the energy market, price signals are also used to simulate network limitations. Finally, utilizing the clustering technique, a segmentation strategy is suggested to improve the scalability of P2P marketplaces.

1.9.3 CONTROL OF AN LV MICROGRID WITH PEER–PEER ENERGY TRADING

This Figure 1.7 illustrates the structure of a four-tier system used to identify and classify key components and techniques involved in figure-to-peer energy trading based on the roles they play.

There are three levels in the system architecture. In the first dimension, P2P energy trading's primary operations are split into four interoperable tiers. Each layer is described in the paragraphs that follow. Feeders, transformers, smart metres, loads, and DERs are all included in the power grid layer (DERs). These components make up the actual power distribution network, which is where P2P energy trading takes place.

The ICT layer is made up of hardware, software, protocols, and data flow. Communication devices include sensors, telecommunication connections, routers, switches, servers, and other types of computers. A few examples of protocols include TCP/IP, PPP, X2.5, and others. Examples of communication applications include data transfer and file sharing. Communication involves the sender, recipient, and content of all messages sent through communication devices.

The control layer primarily comprises the power distribution system's control functions. This layer specifies a few control techniques for maintaining and regulating the quality and dependability of the power supply. For example, voltage, frequency, and active power regulation are all control activities in the control layer.

The business layer defines how power is shared between peers and with third parties. The main players in the energy market are peers, suppliers, DSOs, and the regulatory bodies that oversee the market. P2P energy trading may be implemented using a few business models built on top of this layer.

The second component of the system design is the size of the peers involved in P2P energy trading, such as buildings, microgrids, cells, and regions (consisting of multi-cells).

One single home linked to the electrical distribution system is referred to as an individual premise. DERs and loads are combined in microgrids, which operate in a controlled and coordinated way. It may be connected to the main or separate power grid. There is a medium-voltage/low-voltage (MV/LV) transformer through a few individual buildings and distributed power sources (DERs) in a limited geographical area. It's the same with the concept of microgrids and cells, both of which have been proposed in the study. In a larger area network, it refers to the management of a collection of DERs in response to various objectives. A cell may be grid-connected or islanded, and each microgrid it contains can operate independently.

Even though a region may be smaller than a city, it may include more cells than an actual metropolitan area would have. They may all trade with each other and are all microgrids, cells, or regions.

The third-dimensional timeline depicts the evolution of the P2P energy trading mechanism through time. Energy clients (generators, consumers, and proactive) agree to trade in advance of the energy exchange by bidding on contracts. During the bidding process, energy users communicate with one another to come to an agreement on the exchange price and quantity. To complete the second phase of the energy exchange, energy must be generated, transmitted, and consumed by the end user. Settlement agreements and payment are used throughout the settlement process to handle bills and transactions.

An energy seller who has committed to sell a particular quantity may not be able to generate the exact amount stated in the orders because of physical network constraints and the unpredictability associated with DERs. In a similar way, a customer is in a predicament. The difference between promised and actual energy output or consumption must be calculated and taxed at the settlement.

Equilibrium trading in energy may be achieved using several components and technologies. Identification refers to the process of finding or identifying key components or technology of a P2P energy trade. Categorization is the process of organizing important components and technologies into various levels.

Classification in P2P energy trading is based mostly on the functionality, peer size, and period of the participants.

1.9.4 Energy Management of Different DERs by Using Optimization Algorithms

The following important issues should be addressed in many areas of technology and science, including engineering, physical sciences, and economics, for the electrical DG or energy system:

- Production resources must be used within a certain time frame.
- Interconnection between the systems under consideration.
- The duration of the planned home appliances' operation.

All these issues may be resolved by "optimizing" the EMS in microgrids or DGs. As indicated in Figure 1.8, three types of methods may be used to provide a proper operation and control of the examined systems in the DG: (1) Rule-based techniques, (2) optimization-based strategies, and (3) hybrid techniques.

To begin, in the rule-based approach, reference points are assigned depending on the current condition and the definition of certain scenarios, typically using decision trees. This method adapts to system circumstances by offering viable options, although it cannot guarantee the optimum answer. Second, optimization-based methods aim to offer the most effective local or global solutions. In general, an optimization problem's mathematical formulation is to maximize or reduce an objective function while fulfilling all relevant constraints linked to the model's integrated components.

This approach may be handled using precise or approximate techniques, depending on the complexity and difficulty of solving the system issue. Approximate

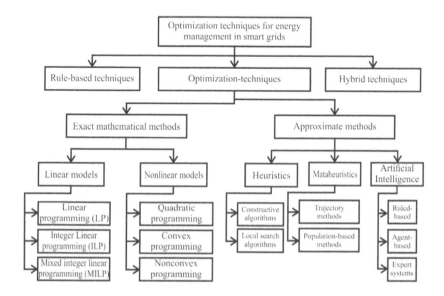

FIGURE 1.8 Classification of optimization techniques for energy management in smart grids.

Practical Implementation of VPP

techniques have the benefit of being able to handle non-linear constraints and goal functions quickly, but they cannot guarantee the quality of the results since they utilize random search methods. Furthermore, as the magnitude of the issue under consideration becomes larger, the likelihood of finding a global solution diminishes.

When precise mathematical techniques are defined in a viable area, they provide an optimum solution. Based on all established restrictions and goal functions, there are two types of models: Linear and non-linear. The linear models are classified into three types: Linear Programming, Integer programming, and mixed-integer linear programming (MILP), depending on whether the variables are real, integer, or both. Finally, hybrid approaches may combine multiple methods to make use of their unique features.

It was decided to use the MILP to model the energy management problem in this chapter because it allows us to make decisions on the operation status of production systems, battery storage systems, EVs, and smart appliances in microgrid smart homes using the characteristics of the integrated DER with integer and binary variables. The MILP and the greedy technique (constructive algorithm) were coupled to create a hybrid methodology that can provide a global near-optimal solution to the prolonged optimization time horizon problem in the Electric System Grid context. In every simulation period, the suggested math-heuristic algorithm produces a local optimum solution in order to get the global near optimal solution in a reasonable amount of time. Previous research and efforts focused on developing different optimization and computational intelligence methods in a DG context as shown in Figure 1.9.

1.10 BLOCKCHAIN TECHNOLOGY

The need for electricity is continuously increasing as the industrial era progresses. To ensure efficient energy distribution, minimum losses, and high quality, as well as the security of power supply, the smart grid concept must be provided. It is possible, as shown in the figure, for individuals to generate and sell energy on a tiny, individual scale. A downside to this concept is that it would complicate how producers and consumers now execute, authenticate, and record transactions in our present system. The network works as a transaction verifier using smart contracts to execute transactions.

Because of the immutability provided by the blockchain, every transaction between producers and consumers is guaranteed to be successful. In addition, it provides transaction history immutability, which may be used in audits or dispute resolution. People can have trust in one another because blockchain technology gives them the ability. The term "blockchain" refers to both a distributed network and an immutable database. In 2008, Satoshi Nakamoto created the bitcoin cryptocurrency with the help of the blockchain technology. Existing blockchain platforms are used by certain cryptocurrencies, while others have their own. One is referred to as a coin, while the other is referred to as a token. As opposed to this, the blockchain isn't limited to cryptocurrencies. Automated transaction settlements are important blockchain implementation concepts in the financial sector in addition to tracking as shown in Figure 1.10.

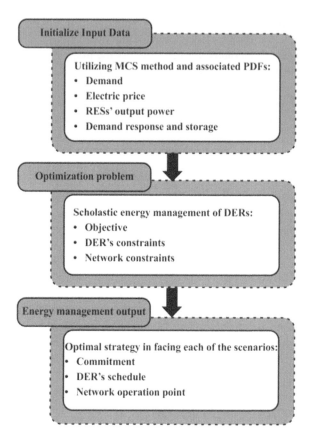

FIGURE 1.9 Deterministic and probabilistic models for energy management in distribution systems.

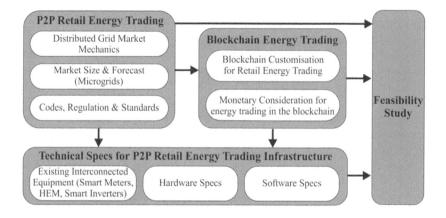

FIGURE 1.10 Blockchain Energy Trading.

1.10.1 Types of Blockchain Techniques

Blockchain, also known as distributed ledger technology, has piqued the attention of companies in the energy industry. Energy firms and technology developers are joining up with governments and universities to use blockchain technology to store energy. The marriage of blockchain and energy is undeniably one of the most exciting technological developments in recent memory.

Trading and crediting are two common blockchain applications in the energy industry. Interested parties use blockchain technology to create a virtual grid that facilitates energy transactions on a distributed or wholesale basis. Consumers may exchange goods and services among their own gadgets and resources, as well as with their neighbours and the grid. The whole procedure may be automated using smart contracts.

One of the most disruptive and interesting cases surrounding blockchain energy is the usage of blockchain in P2P energy trade. The flow of blockchain-based energy trade is shown in Figure 1.11. It incorporates a variety of elements, including financing, community resilience, and renewable energy growth. Renewable energy asset owners (for example, individuals or organizations with solar panels) may exchange excess power with their neighbours if they operate inside a close-knit geographic group or community.

The following are some examples of successful blockchain-based P2P energy trading systems.

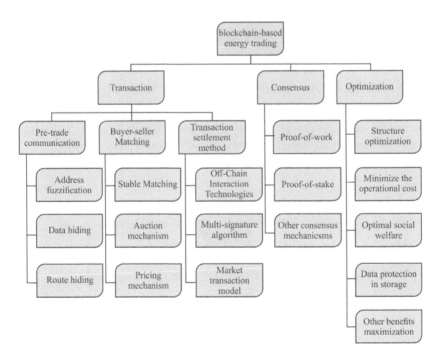

FIGURE 1.11 Flow of blockchain-based energy trading.

- The Brooklyn Microgrid: L03 Energy's Brooklyn Microgrid is a blockchain-based energy initiative that allows individuals to exchange power with their neighbours directly, avoiding big energy utility corporations.

On a general level, homeowners with solar panels use a smartphone app to sell their environmental credit to neighbours without solar panels. Electricity prices may be negotiated directly between buyers and sellers, without the need for middlemen. The blockchain for Distributed Generation facilitates the whole set-up by allowing smart metres to interact with one another consistently.

- ME SOLshare: Through a communal microgrid, ME SOLshare, a blockchain start-up in Bangladesh, is linking houses with solar panels. The company sells the SOL Box, a bidirectional DC energy metre that allows individuals to participate in community trading. Using their built-in wireless internet connection, all SOL Boxes in the network communicate with one another.

A smart contract built on the blockchain connects the quantity of energy produced by a smart metre linked to solar panels to a digital wallet of grid members and banks that provide loans. At the end of the day, banks close their books. It has not only aided in enhancing the participants' livelihoods, but has also aided in extending the productive hours of users, whether for employment or study.

The spread of blockchain technology has expedited P2P energy trading, which has the potential to become a mainstream idea in the decentralized renewable energy industry, benefiting underprivileged people and communities all over the globe.

1.10.2 Microgrid Applications of Blockchain Technology

While there are many benefits to blockchain, the decreased transaction cost is the most important for the energy industry as shown in Figure 1.12. The following are some of the possible advantages of using blockchain in the energy sector:

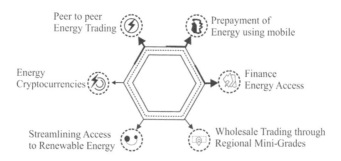

FIGURE 1.12 Application of blockchain in energy.

- Lower utility bills and/or transaction costs in the gas or electricity markets, reducing the demand for operating capital.
- New ways for energy devices to communicate with the grid operator, such as water heaters, electric cars, batteries, and solar PV systems (smart grids). Increased knowledge regarding the integration of variable renewable energy production into the system reduces costs for utilities and grid operators [11].
- Local and decentralized renewable energy networks provide cheap electricity to disadvantaged areas.

1.10.3 DYNAMIC OF ENERGY CONSUMPTION, PRODUCTION, AND EXCHANGE

Over the past few decades, the demand for energy resources and other associated services has grown to satisfy global social demands and economic growth. Numerous companies are engaged in the international exchange of resources to stimulate sustainable growth, and different economies are promoting foreign direct investment and finance production. Environmental degradation and greenhouse gas GHG emissions have risen dramatically because of the rising trend in energy resources. Several studies have shown that the burning of fossil fuels (which accounts for two-thirds of global CO_2 emissions) in support of human activities linked with economic growth and development has been the primary driver of increase in CO_2 emissions. Society is gradually evolving toward more environmentally friendly manufacturing techniques, such as eliminating waste, decreasing automobile air pollution, distributing energy generation, preserving native forests and green areas, and lowering GHG emissions.

Although many factors, including financial development, foreign direct investment, and trade liberalization, are the prerequisites for economic growth, which is the core ambition of every nation. Energy use has become a new paradigm of discussion between economists and practitioners, and this topic has become a focus of serious attention, given the recent progression of the climate agreement. Growing awareness of global warming, unpredictable fossil energy costs, environmental impacts of carbon emissions, and renewable and nuclear energy supplies have emerged as the key pillars of global energy use. Increasing the use of renewable energy and replacing traditional sources with renewable sources is currently an important strategic goal of the economies of many countries. Countries that can create a friendly environment for the development of smart organizations constitute clear examples of a dynamically growing commitment to the effective handling of challenges associated with sustainable development issues. The number of research on sustainable development in management science is growing. Methods and techniques for measuring and comparing the effect of human activities on the environment for various solutions are required for sustainable development. Analyses assessing the impact of natural resources on economic growth have played a key role in creating the analytical framework for sustainable development. Market entities are encouraged to orient their strategies towards sustainable development, which aims at both economic and social well-being on the one hand, equal opportunities for Earth's inhabitants and environmental integrity on the other hand [12].

Moreover, research shows positive attitudes of enterprises towards sustainable development and investments in RESs. Socially responsible activities are beneficial for organizations' brand image and have a positive impact on their innovativeness. Simultaneously, corporate social responsibility offers a complete set of suitable methods for creating value in a sustainable company, which includes fulfilling economic, ecological, and social objectives. Customers are increasingly worried about environmental problems and the long-term consequences of human activities, and their buying behaviour is influenced by ecological features of companies, which they see as green brand equity. Annual data from 1971 to 2009 is used to analyse the connection between India's energy usage, economic growth, and financial development. Adopting responsible energy is strongly influenced by people's ecologically favourable views, environmental awareness, and responsibilities.

According to the results of co-integration using the Automatic Regressive Distributed Log (ARDL) approach, energy consumption is positively and significantly affected by the proportion of urban dwellers in the total population, but it also negatively affects economic growth and the ratio of industrial production to total output as shown in Figure 1.13. Co-integration analysis results show that increasing the urban population ratio has a negative impact on economic growth, while increasing the amount of energy used has a positive impact. As we have found that energy

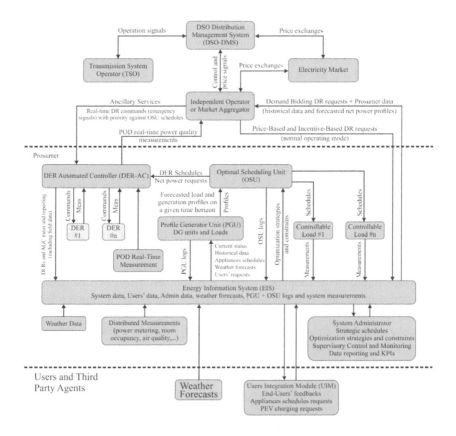

FIGURE 1.13 Automatic Regressive Distributed Log (ARDL) approach.

Practical Implementation of VPP

consumption has a beneficial effect on economic growth, the findings have significant policy implications for the Indian economy. However, it is suggested that India exercises caution when it comes to the efficient use of energy through appropriate channels.

1.11 CASE STUDIES

Because of India's enormous population and fast-growing economy, there is a high need for energy, particularly clean, renewable hydropower. India has only developed around a third of its hydropower potential, according to some estimates.

Entura opened an office in Delhi a decade ago to give India its experience in hydropower design, construction, management, and maintenance.

Since then, Entura has worked as a consultant on both big and small hydropower projects throughout India. It, for example, worked with the developer of the large Chanju-I hydropower project in Himachal Pradesh, which began operations in 2016. Entura employs local engineers, geologists, and other personnel, and its Indian headquarters serves as a hub for project coordination throughout the area, including Nepal and Laos [13].

CASE STUDY: IEEE (BUS ADOPTED IN REAL TIME)

To manage Renewable Energy Resources (RESs) in virtual power plants (VPPs), researchers used the binary particle swarm optimization (BPSO) technique. The use of renewable resources over purchasing from the national grid is essential to keeping costs down. The proposed approach is put to the test using the IEEE 14 bus system, which incorporates main grids and RESs in the form of VPP. Real-world load ordering in Perlis, Malaysia, is used throughout the day to model and reflect system test cases. BPSO also utilizes meteorological data from the Malaysian Meteorological Department to calculate the best ON and OFF times, as well as data on the status of the wind, solar, fuel, and battery. BPSO algorithm reduces VPP's energy use and emissions, according to the results. With this study, a new optimization technique for an main grid-integrated VPP's optimal scheduling controller will be developed to minimize carbon emissions and manage RESs more effectively. An analysis of the most effective algorithms against conventional algorithms shows that integrating RESs improves energy management and reduces emissions. Development of an optimization method is made for regulating power exchange across linked systems with the following goals:

- Reduce the cost of power generation, grid energy purchase, and energy storage
- Incorporate VPP and main grid into a long-term energy management system together with alternative energy sources

- Use optimal VPP scheduling to provide a reliable and high-quality power supply for the loads at the lowest feasible operational cost
- It's important to increase customer confidence by making the system more reliable
- DGs, loads, power flow, and system operational limitations all contribute to power losses
- Reduce CO_2 emissions as planned
- Modelling of VPP and main grid systems

There is an IEEE 14 bus system that provides distribution architecture with real and actual data for loads in each bus and distribution impedance. This study utilizes five main grids to spread out the feed loads across the vehicles. Each main grid has information from five different places. The IEEE 14 bus system, load distribution, and main grid development are all examined in depth in the following report.

1.11.1 IEEE 14 Bus Standard System

Main grid sources and loads are displayed on an IEEE 14 bus in Figure 1.14. In contrast to the previous system, which had two generators to control the grid, this new system only had one generator. A diesel generator, solar panels, a wind turbine, an FC, and a battery are all included in each main grid's construction. Although conventional bus networking systems often use per-unit numbers, this study uses real values to estimate actual power levels for control purposes. A total of 200 MW of power is provided by the main grid, which is linked to the bus and converts 33 kV–11 kV at 50 Hz at the main substation transformer. To improve system dependability, PQ, and transmission line losses, five main grids were placed in various bus bars across the system. In comparison to a single main grid system, a study of the IEEE standard system 1547 shows that many main grid systems have better operating characteristics, making the system more stable and dependable.

To avoid tripping during stand-alone operation, the bus's capacity covers the power provided by each main grid, which totals 10 MW. Buses 5, 6, 10, 11, and 13 all have main grids, as shown in Table 1.4. In total four main grids are used, each of which is identical in terms of size, number of sources, and linkage to other main grids.

The system has nine loads, which are situated at bus 2, bus 3, bus 4, bus 5, bus 6, and bus 14 correspondingly as shown in Figure 1.15. Every bus serves as a feeder to meet the needs of a particular loading location. Perlis, Malaysia, has a real-time load demand curve that may be used as an example.

In a genuine system, generations could undoubtedly decrease energy usage! RESs in the main grids, on the other hand, need a strong controller to effectively arrange their work. For this reason, the aim of this study is to develop a new scheduling controller based on the binary particle swarm optimization (BPSO) algorithm for the IEEE 14 bus power management system, which includes several main grid's

Practical Implementation of VPP

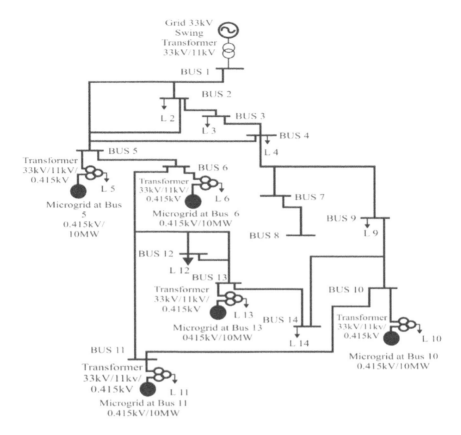

FIGURE 1.14 IEEE 14 bus system.

affiliated with RES to set up a VPP. In February 2016, real-time load data from Perlis, Malaysia, was used to model and simulate the VPP system. According to the suggested BPSO algorithm, the VPP with RESs-integrated main grid's sustainable energy use had the best scheduling control possible. When compared to a random scheduling controller, the obtained findings indicate a substantial contribution to reducing grid energy consumption by 47 per cent and CO_2 emissions by 8.46 per cent. BPSO outperforms other algorithms and controllers in terms of power savings, cost savings, and pollution reduction.

CASE STUDY: IEEE 30 (ADOPTED IN REAL SYSTEM)

The idea of a virtual power plant (VPP) is introduced by combining distributed energy resources (DERs) and proactive into a single plant to gain associated advantages through a market interface (VPP). Techno-economic aggregation is imposed by this aggregation.

VPP operators have difficulties while engaging in the Locational Marginal Price (LMP)-based wholesale market. Furthermore, geographically distributed DERs linked to transmission buses at various nodes suffer numerous LMPs. This would influence VPP scheduling and trading choices, necessitating more research. This chapter provides a mathematical explanation for network-restricted VPP scheduling with LMP-based wholesale market participation in this respect. To implement the suggested paradigm, the IEEE 30 bus system is changed as shown in Figure 1.16. The LMPs greatly assist the VPP operator in allocating resources efficiently while keeping network traffic below acceptable bounds. Furthermore, to optimize profit, power imbalance is exchanged in an LMP-based market via numerous nodes. Considering technological and commercial considerations, the proposed model would assist VPP operators in maintaining a consistent system.

TABLE 1.4
Bus to Bus Impedance for IEEE 14 Bus System

Line No.	Bus No.	Line Impedance		Half-Line Charging Susceptance (p.u.)
		R (p.u.)	X (p.u.)	
1.	1–2	0.01938	0.05917	0.02640
2.	2–3	0.04699	0.19797	0.02190
3.	2–4	0.05811	0.17632	0.01870
4.	1–5	0.05403	0.22304	0.02460
5.	2–5	0.05695	0.17388	0.01700
6.	3–4	0.06701	0.17103	0.01730
7.	4–5	0.01335	0.04211	0.0064
8.	5–6	0.0	0.25202	0.0
9.	4–7	0.0	0.20912	0.0
10.	7–7	0.0	0.17615	0.0
11.	4–9	0.0	0.55618	0.0
12.	7–9	0.0	0.11001	0.0
13.	9–10	0.03181	0.08450	0.0
14.	6–11	0.09498	0.19890	0.0
15.	6–12	0.12291	0.25518	0.0
16.	6–13	0.06615	0.13027	0.0
17.	9–14	0.12711	0.27038	0.0
18.	10–11	0.08205	0.19207	0.0
19.	12–13	0.22092	0.19988	0.0
20.	13–14	0.01709	0.34802	0.0

Practical Implementation of VPP 35

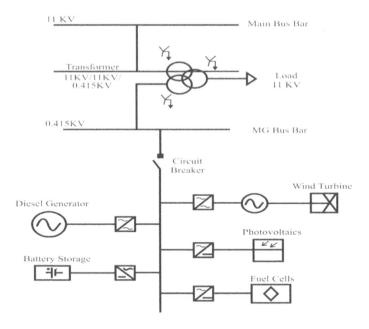

FIGURE 1.15 Single-line diagram or proposed interconnected main grid system.

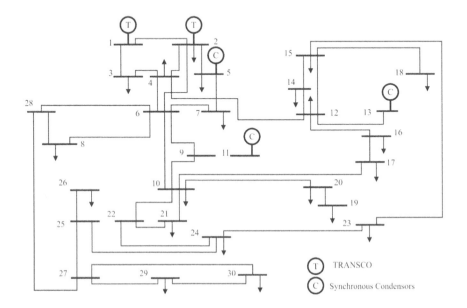

FIGURE 1.16 IEEE 30 bus system.

The company has two main difficulties in providing clean and cheap energy to its customers: Environmental threats and increasing fuel costs. Government laws have changed, making it simpler to deploy DERs like rooftop solar PV on a large scale. However, the intermittent nature of renewable-based DERs has severe implications for grid stability and economic sustainability. The concept of a VPP may be a solution to these issues by coordinating and scheduling DERs, storage, and flexible loads. This could lead to safe operation with high penetration of DER and reduced utility peak-load. This research makes use of the Distributed Energy Resources Customer Adoption Model (DER-CAM), a decision-making tool based on MILP. The possibilities and implications of VPP should be determined via an analysis of Punjab State Power Corporation Limited (PSPCL), a state-owned power company.

The Indian power sector is progressing towards adopting advanced technologies and new regulations. Figure 1.17 illustrates the key developments taking place in India, which are demanding a strengthened and modernized grid, and forming a base for deploying VPP solutions in India.

1.11.2 Existing Policies and Regulations Indirectly Supporting the Adoption of VPP in India

April 1, 2020, is the deadline set for the migration to five-minute scheduling and metered accounting and settlement by the Forum of Regulators' "Introduction of Five-Minute Scheduling" sub-group report (February 2018). When it comes to setting up five-minute bidding and moving to "fast" markets through DERs, India's VPPs may be a helpful instrument.

"Re-designing Real Time Electricity Markets in India" by CERC (July 2018) discusses real-time pricing to create a line of demarcation between "energy trading" and "system imbalance management." Only when energy production from resources

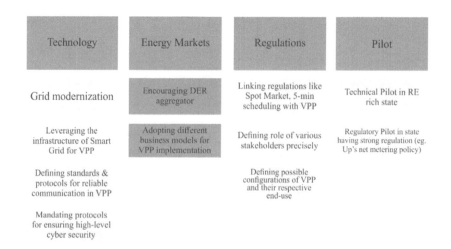

FIGURE 1.17 Advanced technologies and new regulations in India.

is predictable and controllable is a transition to real-time markets feasible. VPPs may help solve this problem by making RE/DER more manageable.

The establishment of a reserve market that may bid at regional and national levels is the subject of CERC's Re-designing Ancillary Services Mechanism in India discussion paper from September 2018. The research also points to DERs as a potential source of supplementary services in the future. DER collection and control and visibility by VPPs may help with this transition and re-design of the market for auxiliary services.

To ensure proper RE generation, forecasting, and scheduling, the CERC's draught regulation on "Deviation Settlement Mechanism and Related Matters 2013" and Latest (Fifth) Amendment, 2019, aims to penalize generators that depart from the scheduled producing schedule. Through improved forecasting and optimization capabilities, as well as the capacity to manage producing sources in real time, VPPs may aid in the reduction of power curtailment and/or deviation costs.

- APERC's regulation on "Solar and Wind Generation Forecasting, Scheduling, and Deviation Settlement Regulation": The state of Andhra Pradesh has made provision for virtual pooling of various pooling stations to take advantage of wider geographical regions and variety. Virtual pooling may be aided by VPP, which performs all the necessary functions to bring generators together.
- UPERC's "Net Metering Policy of Uttar Pradesh": It allows for group net metering and P2P transactions. VPPs may have a major impact on the future acceptance of these technologies. These are important developments in India that demonstrate the importance of VPP in India.
- DERC's 2014 "Demand Side Management Regulations": It describes "Demand Response programmes" as part of load management. In addition, the regulation's definition of Energy Service Companies has left the door open for other parties to accomplish the necessary load decrease. VPP may be used by these third parties as a method to accomplish the resulting load reduction. VPP's many capabilities may also be used to allow demand response programmes to reduce load.

For VPPs, several dependent technological components are crucial for their functionality. The following systems are to be regarded as part of a VPP:

- The central SCADA system as the central control instance
- The remote-control units/control boxes (usually with a programmable logic controller with a communication interface to the power plant, plus a communication unit for remote communication) on site at the plants for the realization of remote metering and remote control
- The communication link via satellite or Internet, where appropriate, as a closed user group (for example, in the case of balancing service provision)
- The database systems for the storage of various data sets, measurement data, schedules, or forecasts with a data management system to manage the data records

- Control centre based on a graphical user interface (GUI), which serves as an interface between the control personnel and the VPP system
- Optional further software components such as an EMS for the plant optimization or forecasting modules that use externally provided information, such as weather forecasts, to create a load/generation or electricity price forecast

REFERENCES

1. Bagchi, A. "Power system adequacy and cost/worth assessment incorporating virtual power plants," PhD Dissertation, Nanyang Technological University, Singapore, 2018. [Abstract available at: https://repository.ntu.edu.sg/handle/10356/73173, full report available from July 1, 2019].
2. Wang, J., Zhong, H., Xia, Q., et al.: "Optimal planning strategy for distributed energy resources considering structural transmission cost allocation," IEEE Trans. Smart Grid, 2018, 9, 5, pp. 5236–5248.
3. Bagchi, L.G., Wang, P.: "Generation adequacy evaluation incorporating an aggregated probabilistic model of active distribution network components and features," IEEE Trans. Smart Grid, 2018, 9, 4, pp. 2667–2680.
4. Kasaei, M.J., Gandomkar, M., Nikoukar, J.: "Optimal management of renewable energy sources by virtual power plant," Renew. Energ., 2017, 114, pp. 1180–1188.
5. Nosratabadi, S.M., Rahmat-Allah, H., Gholipour, E.: "A comprehensive review on microgrid and virtual power plant concepts employed for distributed energy resources scheduling in power systems," Renew. Sustain. Energy Rev., 2017, 67, pp. 341–363.
6. Zubo, R.H.A., et al.: "Operation and planning of distribution networks with integration of renewable distributed generators considering uncertainties: a review," Renew. Sustain. Energy Rev., 2017, 72, pp. 1177–1198.
7. Muttaqi, M., Le, A.D.T., Aghaei, J., et al.: "Optimizing distributed generation parameters through economic feasibility assessment," Appl. Energy, 2016, 165, pp. 893–903.
8. Othman, M.M., et al.: "Optimal placement and sizing of distributed generators in unbalanced distribution systems using supervised big bang–big crunch method," IEEE Trans. Power Syst., 2015, 30, 2, pp. 911–919.
9. Xin, H., et al.: "Virtual power plant-based distributed control strategy for multiple distributed generators," IET Control Theory Appl., 2013, 7, 1, pp. 90–98.
10. Xu, Y., Singh, C.: "Adequacy and economy analysis of distribution systems integrated with electric energy storage and renewable energy resources," IEEE Trans. Power Syst., 2012, 27, 4, pp. 2332–2341.
11. Blockchain, S.M.: Blueprint for a new economy. O'Reilly Media Inc., 2015. [https://www.amazon.in/Blockchain-Melanie-Swa/dp/1491920491].
12. Tapscott, D., Tapscott, A.: "Blockchain revolution: how the technology behind bitcoin is changing money, business, and the world," Penguin, 2016 [https://www.goodreads.com/book/show/25894041-blockchain-revolution, accessed November 20, 2017].
13. Imbault, F., Swiatek, M., De Beaufort, R., Plana, R.: The green blockchain: managing decentralized energy production and consumption. In: Proceedings of the 2017 IEEE International Conference Environ Electr Eng and 2017 IEEE Ind Commer Power Syst Europe (EEEIC/I & CPS Europe), IEEE, 2017, pp. 1–5.

2 Virtual Power Plant with Demand Response Control in Aggregated Distributed Energy Resources of Microgrid

Elutunji Buraimoh and Innocent E. Davidson
Department of Electrical Power Engineering, Durban University of Technology, Durban, South Africa

CONTENTS

2.1 Introduction 39
2.2 Virtual Power Plant in Renewable Energy-Based Microgrids 41
2.3 Virtual Power Plant and Smart Grids...43
    2.3.1 Proposed Implementation of Virtual Power Plant...........................44
    2.3.2 Demand Response Communication Protocols 45
2.4 Control Technique 47
2.5 Electricity Consumption Prediction 49
2.6 Electricity Market's Demand Response 50
2.7 Conclusion 50
References..50

2.1 INTRODUCTION

Power systems face certain technical challenges owing to the rising penetration of distributed renewable energy resources to address environmental issues such as zero-carbon emissions. Similarly, this issue is further complicated with the local control systems of these individual distributed energy resources (DERs). The capacity of a single DER may be insignificant to system operators; however, several DERs in the system result in power over generation beyond the system's capacity, reduced efficiency, and higher costs [1]. Forming virtual power plants (VPPs) is one method to utilize DER capacity and improve controllability entirely. The consolidation of DERs into VPP improves the capacity that may be used. VPPs also allow power system operators to achieve optimal power flow and economic goals [2]. Thus, renewable

DOI: 10.1201/9781003257202-3

energy source (RES) and energy storage system (ESS) have steadily risen in contemporary electrical power networks.

As a result, new ways to regulate them, like microgrids and VPP, have emerged. In general, VPPs are autonomous systems with efficient power flow management systems. VPP is typically a collection of DERs analogous to a traditional power plant and includes RESs, energy storage, and regulated loads. VPP overcomes the limitations of an individual DER's capabilities, increasing its flexibility and controllability. As a result, the VPP may operate and exchange energy the same way as a traditional power plant of equivalent size. In addition, grid services such as frequency and voltage restoration and energy trading are available through the VPP. VPPs have been used in remarkable ways all across the world. However, many issues remain in VPP utilization, including power demand allocation, control system design, and communication networks. VPPs consist of many components connected to the distribution network. However, it is time-consuming and challenging to coordinate the whole VPP [3].

Customers' ability to actively participate in energy transactions is crucial for demand-side management in modern electricity markets. This participation has opened up several opportunities for system operators and utilities to directly influence demand through various price and incentive-based mechanisms to avoid extreme grid conditions or power outages and achieve efficient and cost-effective operation. The demand response (DR) is one method that can significantly impact at a modest direct cost [4]. Peak loads have become a global problem because of rising energy needs since they may create system instability and high electricity prices, resulting in higher end-user electricity bills. The traditional answer is to construct electricity production plants that can handle peak loads, rising fuel prices, and carbon emissions. As a result, reducing demand during peak hours is critical, resulting in a win-win scenario for both utilities and consumers. The utility business does not need to invest in costly additional generation capacity, lowering customer power bills. With the growing use of smart grids, research and practice have demonstrated demand-side involvement and DERs aggregation as microgrids [5]. Microgrid and distributed resources are scattered generating units and storage facilities that promise to reduce peak loads when combined into a unified framework.

A VPP is designed to incorporate DERs, RESs, ESSs, and consumers with controlled loads into a single plant. In general, a VPP engages in two sorts of transactions. On the one hand, it bids in the wholesale market on behalf of enrolled players within its area; on the other hand, it might run its inner market by coordinating the activities of these participants. Furthermore, a VPP allows customers to actively engage in the VPP internal market using a well-designed DR scheme that helps smooth aggregated net load and enhances grid stability [6].

The communication, information, and functional needs of VPPs are evaluated in [7]. First, the prerequisites for interoperability are conceptualized, and a comparison of their fulfilment by state-of-the-art communication systems is offered in [7]. Then, the VPP requirements are translated into IEC 61850 services and information models and Common Information Model (CIM) power utility automation standards. Thus, extensions to the IEC 61850 standard are proposed in [7] to improve the interaction between the VPP controller and DERs. For the stability study of power systems, [8]

Virtual Power Plant with Demand Response Control   **41**

provides an aggregated model to capture VPP transients. The purpose is to develop a model that can be used for system studies and by the transmission system operator to assess the impact of VPPs on the grid as a whole. Study [9] investigates different VPP operating models with the goal of unified administration of many VPP via a VPP central controller, revealing the VPP's controllability as a source and load in general.

In [1], a real-time distributed clustering approach is developed for aggregating dispersed energy storage devices into heterogeneous VPPs. There are two types of VPPs: one for supplying bulk electricity and the other for supplying high-frequency power. Based on the capacity of each distributed ESS and the owner's willingness to join in one of the VPPs, the proposed distributed clustering algorithm in [1] determines the VPP memberships for each distributed ESS. The coordination of dispersed energy resources inside a VPP can help with market and network integration. However, operating a VPP in the face of network constraints and other market and physical variables is a difficult task. This [10] presents a paradigm for co-optimizing the VPP's supply of numerous markets, systems, inertia, and local network services to maximize revenue.

This study discusses the VPP's existing and potential contributions in the context of microgrids. In addition, a detailed description of the VPP for DERs is offered, as well as the architecture and prototype of the accompanying communication platform. Consequently, a DR control model and an electricity consumption model are offered for microgrid users. Finally, the microgrid's interface with the electricity market is investigated.

2.2 VIRTUAL POWER PLANT IN RENEWABLE ENERGY-BASED MICROGRIDS

The more the grid penetration and integration of RESs, the more the grid switches from a generation-oriented to a consumption-oriented system. DR can offer customers more options to facilitate this transition. In this chapter, a VPP is offered as a technique to enable accumulated DR from a renewable energy-based microgrid. The VPP serves as a linkage in the middle of the microgrid's customers and distribution system operators (DSOs), and its course of action must advance the course of both parties. The firms responsible for distributing and managing energy from generating sources to ultimate users are known as DSO. Smart meters and real-time communication are used in the DSO concept, allowing bidirectional energy flow reading and communication. The VPP networks with various investors and industry players in this scenario, and interoperation issues might occur. Furthermore, the VPP controls certain DERs installed in the microgrids and uses sensitive customer data. As a result, it is critical to guarantee that the system is safe and that individuals' privacy concerns are adequately addressed.

The VPP procedure reflects a dual optimization objective in a microgrid: satisfactory control of the local microgrid energy supplies and offering effective DR to benefit both the microgrid customers and DSOs. These control problems are difficult to solve since they involve maintaining customer comfort, using the microgrid's thermal inertia, and dealing with disruptions such as external weather conditions, among other things. Furthermore, accurate electricity consumption forecast models must enable the VPP to function. Owing to individual differences in behavior, achieving acceptable precision and correctness in these models is difficult for a particular customer within a microgrid; however, the work becomes more manageable when combining a class of customers [11, 12]. Finally, the connection with electricity markets is critical for the effectiveness and technical acceptance of VPP's proposed use in microgrids. The VPP can provide these markets with variable load consumption, system energy storage, and DER's output energy from the microgrid. However, combining this flexibility with promises is difficult because some depend on user behavior. Therefore, all of the issues above should be considered when designing a VPP for a microgrid structure.

This chapter proposes a VPP architecture suitable for further research and advanced work to demonstrate the significances of intelligent systems, prediction systems, control systems, and market systems in VPP. Thus, Figure 2.1 depicts the intended VPP's context diagram and interactions with the various actors. A VPP should aggregate decentralized medium-scale power sources, such as solar and wind power plants, combined heat and power units, demand-responsive loads, and storage systems with a double goal. Thus, on the one hand, VPP alleviates the external smart grid's stability and dispatchability issues since it may be operated on an individual basis while appearing as a single system overall. VPP, on the other hand, increases the flexibility of all networked units, allowing traders to improve renewable energy forecasts and trading programs.

In the proposed architecture, RESs, various microgrid local loads and resident energy storage are treated as DER in the microgrid. The microgrid consumers utilize these DERs supervised by an array of sensors and managed by microgrid automatic controllers. The microgrid monitoring equipment also measures nonelectrical

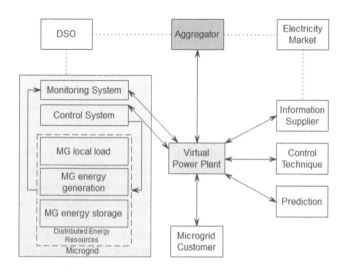

FIGURE 2.1 Proposed virtual power plant.

Virtual Power Plant with Demand Response Control

data such as climatic conditions and carbon emissions. In addition, the VPP gathers information from other sources such as power pricing and grid condition. The VPP's intelligence digests all of this information to govern energy resources using predictive modelling and control algorithms. Thus, predictive modelling and control algorithms are integral parts of the VPP system, and they are placed within the VPP. Predictive modelling is a statistical approach that uses machine learning and data mining to forecast and predict possible future occurrences with past and existing data. Predictive models make assumptions based on what has occurred and is now occurring. A controller's control algorithm is a form of logic that analyzes the difference between a measured value and a set point. This logic aids the controller in processing field input and making decisions. Conclusively, the VPP interrelates with different microgrid customers by offering services or imposing consumption performance limits and with central microgrid operators by trading DR bids. However, in the present-day grid systems, the role of the aggregator is still unclear. The following parts delve deeper into each of the project's examined areas.

2.3 VIRTUAL POWER PLANT AND SMART GRIDS

The development of smart grids and VPPs to dispatch and regulate the flow of electricity into our communities is at the forefront of an energy revolution. The electrical power infrastructure has been for almost a century, but digital technology makes it more intelligent and relieves pressure on an aging grid that more people are using. The smart grid can perform all of the functions of the traditional grid but at a much faster rate. It can also adapt more quickly to changing consumer demand throughout the day and night. The electricity grid in most regions is made up of one-way transmission lines that go from power plants to transformers and substations. If a critical component breaks, it might result in a chain reaction of power outages. This may be avoided with a smart grid that responds fast. In addition, because it is a two-way system, it may compensate for electricity shortages by tapping renewable energy supplies located around the region.

The smart grid will be consolidated in the future, and new smart power technologies will be used entirely in VPPs. On the generating and storage side, these energy hubs allow the system operator to view and control grid activity flowing in and out—to tap into RESs such as rooftop solar panels and electric automobiles while monitoring smart meters and even smart appliances on the consumer side. In addition, energy may be provided from DERs while the load on the grid is balanced by remotely shifting customer consumption to efficiency mode using digital communication technology. In a smart grid scenario, VPPs bring up a new range of benefits while also posing significant hurdles to overcome. VPPs are viewed as a means of integrating DERs into the power grid and the energy market. A VPP must gather and digest data from various sources and deliver control signals to the DER in compliance with regulations, posing interoperability difficulties that must be resolved through the use of standards. In addition, the VPP must communicate with customers, delivering services and allowing them to customize operation parameters.

The architecture for VPP is discussed in [13] and the interaction of customers' meters through the VPP controller. Similarly, the article creates an human-machine interface (HMI) to access reactive power metering at the customer's end and a recording tool for VPP controller values. The communication, information, and functional needs of VPPs are evaluated in [7]. First, a conceptual definition of the interoperability requirements is offered, followed by comparing how state-of-the-art communication mechanisms meet them. Next, the VPP requirements are translated to IEC 61850 services and information models and CIM power utility automation standards. Finally, changes to the IEC 61850 standard are proposed to improve the interaction between the VPP controller and DERs. For the management of VPPs, [14] presents a distributed decentralized prediction approach. The suggested approach successfully merged neural networks and machine learning methods with a distributed architecture appropriate for VPP aggregation. VPPs have also piqued the interest of large corporations such as Siemens, General Electrics, and Schneider Electric, who are creating and developing VPPs for their clients.

The VPP is designed in Table 2.1, the suggested is developed for microgrids, and the design is separated into six distinct components. First, the monitoring system collects data from various data sources via various protocols and transfers it to the data storage system. The data storage system performs all gathered data activities with an integrated microgrid database for near real-time monitoring and control. Additionally, this module transfers all captured data to an external data storeroom for historical data analysis. Third, the control system acts as the VPP's brain, regulating and responding to demand utilizing direct and indirect control. The VPP also employs prediction models such as power usage and presence detection within the prediction system. Fourth, the bidding system manages the interface with the aggregator earlier shown. Finally, the support functions system of the VPP encapsulates other tasks such as data validation and consumer engagement [15].

2.3.1 Proposed Implementation of Virtual Power Plant

The suggested VPP is intended as a microgrid's energy management system (EMS) with flexibility and accommodating microgrid expansion. The project can be implemented by computational modelling and simulation or prototyping because of hardware procurement financial constraints. The prototype system will deploy a sensor network employing industrial automation technology. A programmable logic

TABLE 2.1
Microgrid VPP Component

	Components	Functions and Applications
1	Monitoring	Price of electricity, weather forecast, and sensor data
2	Controlling	Drivers, constraints, and direct and indirect controls
3	Data storage	Microgrid database, data access, and external data storeroom
4	Prediction	Consumption of electricity and presence detection
5	Support functions	Validation of data and consumer interaction
6	Bidding	Cost calculation of demand response, bid calculation, and handling of requests

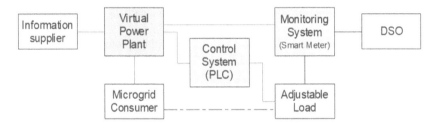

FIGURE 2.2 Implementation of the proposed architecture.

controller (PLC) must be installed in all the microgrid units to collect data from all sensors. In addition, for interoperability considerations, an embedded computer will be used to host the VPP. Figure 2.2 depicts the implementation of the proposed VPP architecture.

External information suppliers supply a microgrid VPP with grid CO_2 emissions forecasts and day-in-advance power pricing. In addition, the VPP is proposed to use any meter protocol to collect power usage datalog from the smart meter. For example, the microgrid customer can schedule laundry activity (with a start and stop time) using the washing machine and choose among a green program (CO_2), an economic program (electricity price), or a well-organized combination of the two; using an internet-based user interface. The VPP's control algorithm determines when the washing machine should be operated; it sends a control signal to the control system (PLC) through the automation device specification (ADS) protocol [16]. ADS will allow data to flow across various program components. Conclusively, the PLC initiates the washing machine with a pulse.

2.3.2 Demand Response Communication Protocols

The presence and application of standards is critical enabler for developing DR and smart grids. The usage of standards ensures interoperability across various stakeholders. Several smart grid standards have arisen, and various organizations in various countries are developing them. The National Institute of Standards and Technology (NIST) in the United States is in charge of standardization, whereas European Committee for Electrotechnical Standardization (CEN-ELEC) in the European Union is in charge [17]. Standard protocols are required to construct solid business cases around emerging issues such as DR from a communication standpoint.

Selecting a proper set of protocols to guarantee a smooth, secure, and interoperable system for information and communications technology (ICT) service rollout for the smart grid is a major success element for the future. One factor that prevents horizontal integration of microgrid automation services across multiple application domains such as microgrid units and smart grid services is interoperability for the local area network [18]. The usage of open protocols specified by the Internet Society appears to be a recent trend in protocol specifications for the smart grid. The Internet Engineering Task Force (IETF) has issued RFC 6272 that outlines a set of internet protocols that may be utilized with the smart grid [19]. This standard supersets proposed protocols that do not include configuration specifics for individual smart grid applications.

Smart grid systems use different and incompatible protocols; Microrequest-Based Aggregation, Forecasting, and Scheduling of Energy Demand, Supply, and Distribution (MIRABEL) and Open Automated Demand Response (OpenADR) are two well-known examples of intelligent DR systems [20]. Similarly, there is ongoing work on DR standardization, automation, and simplicity. For example, the OpenADR Alliance was formed to standardize, automate, and simplify DR and DER, allowing utilities and aggregators to manage and expand energy demand and decentralized energy generation more cost-effectively, enabling customers to have more influence over their energy destiny. OpenADR is a two-way information exchange model and smart grid standard that is open, highly secure, and two-way. We are presently co-creating the future of smart grid upgrading [21]. The OpenADR Alliance created OpenADR 2.0, an application layer standard. The DR automation server is at the heart of the OpenADR protocol (DRAS). The utility side of the DRAS includes things like establishing programs, verifying capacity, and calling events, among other things. The participant or client is on the opposite side, opting in or out of programs, gathering event information. Every OpenADR function is categorized as either utility or participant/client on one side of the process. In many circumstances, each side has similar functions; for example, a client may submit a bid (which is part of the "Automate Bidding" category), while the utility may retrieve all bids when the bidding window ends (which is also part of the "Automate Bidding" category) [22].

The "traditional" ZigBee standard, i.e., versions 1 and 2 of the ZigBee specification, offers a network layer that runs on top of the IEEE 802.15.4 radio media access control (MAC). The protocol stack is an open standard that companies may download and use for free. The ZigBee network layer (NWK) is incompatible with IP and does not connect heterogeneous networks. The NWK protocol is less appealing for the smart grid local area network due to its lack of support for heterogeneous networks. The ZigBee and HomePlug® Alliance has released a definition that includes a number of protocols for the local area network. The ZigBee IP standard is the outcome of this decision [23]. The ZigBee IP standard defines an interoperable stack of IETF-established protocols for use in IEEE 802.15.4 wireless mesh networks. In contrast to RFC 6272, the ZigBee IP standard includes information on how to configure the open protocols that will be utilized. The NWK layer has been replaced with IPv6 across low-power wireless personal area networks, which is a fundamental distinction between old ZigBee and ZigBee IP (6LoWPAN).

This chapter presents a unique approach for evaluating the efficacy of DR protocols and a DR strategy. This approach may evaluate any strategy in conjunction with a given protocol and then repetitively modify parameters in either the strategy or the protocol to obtain desired results. First, using existing specifications, a model is created by formalizing a microgrid scenario description (earlier specified in plain language), a DR strategy, and a protocol. Second, the resulting model is converted into workable code, and the given situation is simulated using the strategy and protocol. Lastly, the outcomes are assessed using a set of performance criteria that apply to both the protocol and the strategy. Monitoring performance indicators allows for fine-tuning of protocol and DR plan parameters.

2.4 CONTROL TECHNIQUE

The VPP's control module intends to optimize the microgrid's local energy supplies while also offering grid flexibility through DR. By altering the load profile and microgrid, automation systems can reduce the cost of power use. This may be accomplished through making a better use of local storage, local energy generation, and flexible loads. Information about the present and future weather situations, power costs, and microgrid conditions is required for this purpose. The microgrid's customer comfort must not be jeopardized during the VPP operation. The presence of external disturbances, such as weather conditions, must also be factored into the equation. Last but not least, the behavior and presence of the sensitive microgrid's consumers must be taken into consideration.

For the VPP's problem, there are primarily two control approaches: direct and indirect control techniques. Direct control is generally provided by a centralized EMS that adjusts DER automatically following a consumer agreement. Indirect control entails supplying customers with information such as power pricing or grid CO_2 emissions to influence their electricity use. Direct control is the most common control method for a VPP; however, a mix of direct and indirect control has been shown to be helpful [24]. As a result, the suggested VPP considers both direct control and indirect control, but only the former is discussed here.

Logic-based EMS are extensively utilized in the literature [25–27]. These systems use tree rules to make decisions (e.g., if-then-else). These systems have a major flaw: they rely on reactive control that means that control actions are done based on the present situation. On the other hand, predictive control examines both present and future situations, which typically results in superior solutions. Model predictive control (MPC) has been utilized in a variety of applications and for challenges that are comparable to those faced by the VPP. Several types of research have been conducted on regulating heating, ventilation, and air conditioning (HVAC) systems, local energy production, and microgrid lighting facilities [28–32]. MPC has also been studied for a microgrid's local storage, production, and loading [33, 34]. One of MPC's key advantages is its capacity to deal with complicated dynamics and system restrictions.

The VPP's suggested control approach is based on the MPC. The DERs in the microgrid will be regulated by this controller, including energy generation, energy storage, and various categories of loads (adjustable and sheddable). The non-sheddable loads will be handled by indirect control. The direct control will be implemented by solving a multi-objective optimization problem that aims to maximize microgrid's customer comfort, reduce power costs, and offer demand responsiveness. The difference between baseline power usage and actual electricity consumption with a DR event will be minimized for this last optimization parameter. Conflicting goals can arise when multi-objective functions are used in complicated systems. In order to solve this problem in an MPC framework, soft constraints might be used to expand the solution space. Because each soft-constraint introduces a new variable into the optimization function, adding soft-constrains increases the problem's complexity. The MPC horizon's length and each time step's resolution are two important factors to consider.

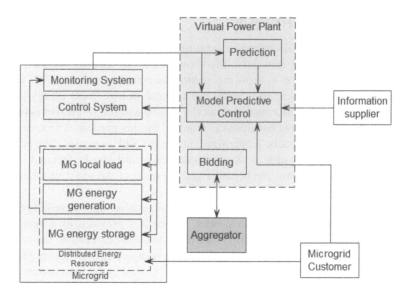

FIGURE 2.3 Virtual power plant with model predictive control for the microgrid.

The suggested MPC in the VPP architecture is shown in Figure 2.3. External information suppliers offer the MPC information on power pricing and weather forecasts. The bidding module communicates with the aggregator and sends a prediction curve of the intended power load and a DR event to the MPC. The microgrid local consumers interact with the microgrid network, the DER within it, and the MPC by placing comfort limitations on it. The VPP's prediction module uses forecasting algorithms to forecast baseline power usage. The MPC optimizes under the following parameters listed and transmits control signals to the microgrid's local controlling system that actuates the various DER. Finally, the monitoring infrastructure completes the control loop by returning to the controller measurements on the DER's and the microgrid's conditions.

The optimization problem of the VPP controller must solve the following by minimizing them

1. Product of the electricity price and base power consumption
2. Microgrid customer's discomfort depending on climatic factors (temperature, humidity, and CO_2 levels)
3. Power consumption with DR provision subtracted from base power consumption

The VPP's proposed control algorithm is based on an MPC. The MPC is based on an iterative process in which a construction model is optimized over a finite time period. Some issues, like the system's nonlinearities and the inclusion of binary variables, should be considered while implementing the MPC.

2.5 ELECTRICITY CONSUMPTION PREDICTION

In the United States, DR systems are already in place for commercial and industrial customers, but adoption in the residential sector is still in its early stages. Estimating normal or baseline power consumption without a DR event is a key difficulty in these programs. The difference in electrical load between with and without DR can be used to assess the customer's monetary return. There are a variety of current approaches, ranging from prior day matching to regression algorithms. There are a variety of approaches for estimating the baseline, but none of them have been proved to be exact enough at the microgrid level [35, 36].

Several microgrid research studies have used machine learning approaches such as support vector regression, random forest, Bayesian networks, and neural networks [37, 38]. Some techniques yield promising results; however, because some utilities cannot employ black-box models, it is doubtful that they will be used in the actual world [39]. Furthermore, many of the recommended solutions are based on data from the United States, where temperature and power use are highly correlated due to air conditioners. However, there are few options in chilly nations where air conditioners are scarce.

There are regularly used factors in the literature, such as the hour of the day, the day of the week, or past intake, among the selected variables. Other data, such as data from sensors on climatic conditions and water use (representing typical human activity), external weather conditions from the prediction model, and grid CO_2 emissions, must be considered. According to [40], the response variable's electricity usage data has been changed by one hour. Therefore, the appropriate power consumption output-input into the model for each of the predictors given to the model is the electricity consumption in the building one hour into the future.

The performance indices can be deplored before analyzing different models: root mean square error (RMSE), mean square error (MSE), mean absolute error (MAE), coefficient of correlation (R), coefficient of variation (CV), and mean absolute percentage error (MAPE) [41]. The statistics MAPE and CV are often employed in the literature to evaluate power consumption prediction models [42–44]. The final model's selection criteria included the model's complexity (i.e., variables employed, number of trees in a random forest and bagging trees, K-nearest neighbors, random forests, artificial neural networks, and logistic regression and hidden layers in neural networks), transparency (i.e., if the model is a black-box model), and prediction accuracy (MSE, MAPE, and CV values).

A study is proposed to compare several strategies for predicting power use and determine which was most appropriate based on model complexity, transparency, and accuracy. Transparency has been given a high priority in order to meet the business demands of DR, leading to the rejection of more accurate but nontransparent models. From a VPP standpoint, the model's openness is less significant than its correctness and complexity. Future research in this area might consider a bigger dataset and additional approaches such as support vector regression or autoregressive integrated moving averages models.

2.6 ELECTRICITY MARKET'S DEMAND RESPONSE

More flexibility on the consumption side is necessary as DER penetration increases in the electrical grid. The potential flexibility of a DER installed in a residential structure might be a valuable commodity in the power markets. A VPP can place a DR bid with a market aggregator. Nevertheless, in microgrid parlance, this flexibility could be limited by the customer's actions. The literature has looked at whether having a flexible consumer in the day-ahead and controlling electricity markets is feasible. The literature has also created and tested a newly regulated electricity market to encourage the integration of small-scale DER. A VPP in this new environment must be explored to assume the current market structure or to build a new market that coexists with the old ones. Similarly, the literature suggests a bidding approach for DR and VPP-provided generating units. However, DR bidding's flexibility potential for a VPP of a microgrid to an aggregator has yet to be investigated.

2.7 CONCLUSION

Microgrid, like VPP, is a method for integrating distributed generation (DG) sources and energy storage into the smart grid. They are developed with sufficient care for the grid-connected and island operating mode. However, information technologies have a problem in managing these VPP. They are constructed and managed by various organizations with varying goals, and they adhere to a variety of control and communication standards. The capacity of a load to adapt to short-term fluctuations in power pricing is known as DR. It is becoming more crucial in managing short-term supply and demand, particularly during peak hours and when coping with changes in renewable energy supplies. However, in prior studies on VPP management, the use of the DR paradigm has not been well studied. Thus, the current and future contributions of the VPP in the context of microgrids are discussed in this paper. In addition, a thorough overview of VPP for DERs is provided, as well as the associated communication platform architecture and prototype. As a result, a DR control model and an electrical consumption model for microgrid consumers are presented. Finally, the microgrid's interface with the electrical market through DR bidding is explored.

REFERENCES

1. R. Zhang and B. Hredzak, "Distributed Dynamic Clustering Algorithm for Formation of Heterogeneous Virtual Power Plants Based on Power Requirements," *IEEE Trans. Smart Grid*, vol. 12, no. 1, pp. 192–204, 2021, doi: 10.1109/TSG.2020.3020163.
2. D. Pudjianto, C. Ramsay, and G. Strbac, "Virtual Power Plant and System Integration of Distributed Energy Resources," *IET Renew. Power Gener.*, vol. 1, no. 1, pp. 10–16, 2007, doi: 10.1049/IET-RPG:20060023.
3. M. J. Nski *et al.*, "A Case Study on Power Quality in a Virtual Power Plant: Long Term Assessment and Global Index Application," *Energies*, vol. 13, no. 24, 2020, doi: 10.3390/en13246578.
4. A. Mnatsakanyan and S. W. Kennedy, "A Novel Demand Response Model with an Application for a Virtual Power Plant," *IEEE Trans. Smart Grid*, vol. 6, no. 1, pp. 230–237, 2015, doi: 10.1109/TSG.2014.2339213.

5. E. Buraimoh, I. E. Davidson, and F. Martinez-Rodrigo, "Decentralized Fast Delayed Signal Cancelation Secondary Control for Low Voltage Ride-Through Application in Grid Supporting Grid Feeding Microgrid," *Front. Energy Res.*, vol. 9, pp. 1–16, Apr. 2021, doi: 10.3389/fenrg.2021.643920.

6. Z. Luo, S. H. Hong, and Y. M. Ding, "A Data Mining-Driven Incentive-Based Demand Response Scheme for a Virtual Power Plant," *Appl. Energy*, vol. 239,, pp. 549–559, Oct. 2019, doi: 10.1016/j.apenergy.2019.01.142.

7. N. Etherden, V. Vyatkin, and M. H. J. Bollen, "Virtual Power Plant for Grid Services Using IEC 61850," *IEEE Trans. Ind. Inf.*, vol. 12, no. 1, pp. 437–447, Feb. 2016, doi: 10.1109/TII.2015.2414354.

8. J. Chen, M. Liu, and F. Milano, "Aggregated Model of Virtual Power Plants for Transient Frequency and Voltage Stability Analysis," *IEEE Trans. Power Syst.*, vol. 36, no. 5, pp. 4366–4375, 2021, doi: 10.1109/TPWRS.2021.3063280.

9. Y. Wang, X. Ai, Z. Tan, L. Yan, and S. Liu, "Interactive Dispatch Modes and Bidding Strategy of Multiple Virtual Power Plants Based on Demand Response and Game Theory," *IEEE Trans. Smart Grid*, vol. 7, no. 1, pp. 510–519, 2016, doi: 10.1109/TSG.2015.2409121.

10. J. Naughton, S. Member, H. Wang, and M. Cantoni, "Co-Optimizing Virtual Power Plant Services Under Uncertainty : A Robust Scheduling and Receding Horizon Dispatch Approach," vol. 36, no. 5, pp. 3960–3972, 2021.

11. X. Wang, K. Li, X. Gao, F. Wang, and Z. Mi, "Customer Baseline Load Bias Estimation Method of Incentive-Based Demand Response Based on CONTROL Group Matching," *2nd IEEE Conf. Energy Internet Energy Syst. Integr. EI2 2018 – Proc.*, Dec. 2018, doi: 10.1109/EI2.2018.8582122.

12. I. A. Sajjad, G. Chicco, and R. Napoli, "Definitions of Demand Flexibility for Aggregate Residential Loads," *IEEE Trans. Smart Grid*, vol. 7, no. 6, pp. 2633–2643, Nov. 2016, doi: 10.1109/TSG.2016.2522961.

13. J. Ali, S. Massucco, F. Silvestro, and A. Vinci, "Participation of Customers to Virtual Power Plants for Reactive Power Provision," *Proc. – 2018 53rd Int. Univ. Power Eng. Conf. UPEC 2018*, Nov. 2018, doi: 10.1109/UPEC.2018.8541978.

14. A. Rosato, M. Panella, R. Araneo, and A. Andreotti, "A Neural Network Based Prediction System of Distributed Generation for the Management of Microgrids," *IEEE Trans. Ind. Appl.*, vol. 55, no. 6, pp. 7092–7102, Nov. 2019, doi: 10.1109/TIA.2019.2916758.

15. S. Rotger-Griful, "Virtual Power Plant for Residential Demand Response," Tech. Rep. ECE-TR-22, Dep. Eng. Electr. Comput. Eng. Aarhus Univ., no. October, pp. 1–33, 2014.

16. H. Dibowski, J. Ploennigs, and M. Wollschlaeger, "Semantic Device and System Modeling for Automation Systems and Sensor Networks," *IEEE Trans. Ind. Inf.*, vol. 14, no. 4, pp. 1298–1311, Apr. 2018, doi: 10.1109/TII.2018.2796861.

17. K. C. Ruland, J. Sassmannshausen, K. Waedt, and N. Zivic, "Smart Grid Security – An Overview of Standards and Guidelines," *Elektrotech. Inftech.*, vol. 134, pp. 19–25, 2017, doi: 10.1007/s00502-017-0472-8.

18. R. Hylsberg, J. Søren, and A. Mikkelsen, "Infrastructure for Intelligent Automation Services in the Smart Grid," *Wirel. Pers. Commun.*, vol. 76, pp. 125–147, 2014, doi: 10.1007/s11277-014-1682-6.

19. C. Alcaraz, R. Roman, P. Najera, and J. Lopez, "Security of Industrial Sensor Network-Based Remote Substations in the Context of the Internet of Things," *Ad Hoc Networks*, vol. 11, no. 3, pp. 1091–1104, May 2013, doi: 10.1016/J.ADHOC.2012.12.001.

20. S. Gökay, M. C. Beutel, H. Ketabdar, and K. H. Krempels, "Connecting Smart Grid Protocol Standards: A Mapping Model Between Commonly-Used Demand-Response Protocols OpenADR and MIRABEL," *SMARTGREENS 2015 – 4th Int. Conf. Smart Cities Green ICT Syst. Proc.*, pp. 382–387, 2015, doi: 10.5220/0005486503820387.

21. J. I. G. Alonso *et al.*, "Flexibility Services Based on OpenADR Protocol for DSO Level," *Sensors*, vol. 20, no. 21, p. 6266, Nov. 2020, doi: 10.3390/S20216266.
22. M. Kolenc *et al.*, "Virtual Power Plant Architecture Using OpenADR 2.0b for Dynamic Charging of Automated Guided Vehicles," *Int. J. Electr. Power Energy Syst.*, vol. 104, pp. 370–382, Jan. 2019, doi: 10.1016/J.IJEPES.2018.07.032.
23. P. Yao, G. Liu, and Y. Liu, "Smart Meter Based on Smart Energy Profile for ZigBee and Power Line Communication Connectivity-Enabled Home Area Network," *Appl. Mech. Mater.*, vol. 599–601, pp. 1696–1699, 2014, doi: 10.4028/WWW.SCIENTIFIC.NET/AMM.599-601.1696.
24. T. Morstyn, N. Farrell, S. J. Darby, and M. D. McCulloch, "Using Peer-to-Peer Energy-Trading Platforms to Incentivize Prosumers to form Federated Power Plants," *Nat. Energy*, vol. 3, no. 2, pp. 94–101, Feb. 2018, doi: 10.1038/s41560-017-0075-y.
25. F. J. Vivas *et al.*, "Multi-Objective Fuzzy Logic-Based Energy Management System for Microgrids with Battery and Hydrogen Energy Storage System," *Electron.*, vol. 9, no. 7, pp. 1–25, 2020, doi: 10.3390/electronics9071074.
26. D. Arcos-Aviles, J. Pascual, L. Marroyo, P. Sanchis, and F. Guinjoan, "Fuzzy Logic-Based Energy Management System Design for Residential Grid-Connected Microgrids," *IEEE Trans. Smart Grid*, vol. 9, no. 2, pp. 530–543, 2018, doi: 10.1109/TSG.2016.2555245.
27. M. Jafari, Z. Malekjamshidi, J. Zhu, and M. H. Khooban, "A Novel Predictive Fuzzy Logic-Based Energy Management System for Grid-Connected and Off-Grid Operation of Residential Smart Microgrids," *IEEE J. Emerg. Sel. Top. Power Electron.*, vol. 8, no. 2, pp. 1391–1404, Jun. 2020, doi: 10.1109/JESTPE.2018.2882509.
28. H. Gong, E. S. Jones, R. Alden, A. G. Frye, D. Colliver, and D. M. Ionel, "Virtual Power Plant Control for Large Residential Communities Using HVAC Systems for Energy Storage," *IEEE Trans. Ind. Appl.*, pp. 622–633, 2021, doi: 10.1109/TIA.2021.3120971.
29. X. Wang, L. Liang, X. Zhang, and H. Sun, "Distributed Real-Time Temperature and Energy Control of Energy Efficient Buildings via Geothermal Heat Pumps," *CSEE J. Power Energy Syst.*, 2021, doi: 10.17775/CSEEJPES.2020.05840.
30. D. T. Nguyen, H. T. Nguyen, and L. B. Le, "Coordinated Dispatch of Renewable Energy Sources and HVAC Load Using Stochastic Programming," 2014 IEEE Int. Conf. Smart Grid Commun. SmartGridComm 2014, pp. 139–144, Jan. 2015, doi: 10.1109/SMARTGRIDCOMM.2014.7007636.
31. H. Liang and J. Ma, "Data-Driven Resource Planning for Virtual Power Plant Integrating Demand Response Customer Selection and Storage," *IEEE Trans. Ind. Inf.*, 2021, doi: 10.1109/TII.2021.3068402.
32. M. Gough *et al.*, "Optimal Scheduling of Commercial Demand Response by Technical Virtual Power Plants," pp. 1–6, Sep. 2021, doi: 10.1109/SEST50973.2021.9543463.
33. T. T. Nguyen, H. J. Yoo, and H. M. Kim, "Applying Model Predictive Control to SMES System in Microgrids for Eddy Current Losses Reduction," *IEEE Trans. Appl. Supercond.*, vol. 26, no. 4, Jun. 2016, doi: 10.1109/TASC.2016.2524511.
34. L. K. Gan, P. Zhang, J. Lee, M. A. Osborne, and D. A. Howey, "Data-Driven Energy Management System with Gaussian Process Forecasting and MPC for Interconnected Microgrids," *IEEE Trans. Sustain. Energy*, vol. 12, no. 1, pp. 695–704, Jan. 2021, doi: 10.1109/TSTE.2020.3017224.
35. T. Samad, E. Koch, and P. Stluka, "Automated Demand Response for Smart Buildings and Microgrids: The State of the Practice and Research Challenges," *Proc. IEEE*, vol. 104, no. 4, pp. 726–744, Apr. 2016, doi: 10.1109/JPROC.2016.2520639.
36. C. Wang, D. Chatterjee, M. Robinson, L. Zhao, J. Li, and R. Merring, "Pricing of Emergency and Demand Response Resources," *IEEE Power Energy Soc. Gen. Meet.*, Nov. 2016, doi: 10.1109/PESGM.2016.7741762.

37. M. Al Karim, J. Currie, and T. T. Lie, "Dynamic Event Detection Using a Distributed Feature Selection Based Machine Learning Approach in a Self-Healing Microgrid," *IEEE Trans. Power Syst.*, vol. 33, no. 5, pp. 4706–4718, Sep. 2018, doi: 10.1109/TPWRS.2018.2812768.
38. M. Elsayed, M. Erol-Kantarci, B. Kantarci, L. Wu, and J. Li, "Low-Latency Communications for Community Resilience Microgrids: A Reinforcement Learning Approach," *IEEE Trans. Smart Grid*, vol. 11, no. 2, pp. 1091–1099, Mar. 2020, doi: 10.1109/TSG.2019.2931753.
39. J. Faraji, A. Ketabi, H. Hashemi-Dezaki, M. Shafie-Khah, and J. P. S. Catalao, "Optimal Day-Ahead Self-Scheduling and Operation of Prosumer Microgrids Using Hybrid Machine Learning-Based Weather and Load Forecasting," *IEEE Access*, vol. 8, pp. 157284–157305, 2020, doi: 10.1109/ACCESS.2020.3019562.
40. X. Zhang, G. Hug, J. Z. Kolter, and I. Harjunkoski, "Demand Response of Ancillary Service from Industrial Loads Coordinated with Energy Storage," *IEEE Trans. Power Syst.*, vol. 33, no. 1, pp. 951–961, Jan. 2018, doi: 10.1109/TPWRS.2017.2704524.
41. T. Ahmad and H. Chen, "Short and Medium-Term Forecasting of Cooling and Heating Load Demand in Building Environment with Data-Mining Based Approaches," *Energy Build.*, vol. 166, pp. 460–476, May 2018, doi: 10.1016/J.ENBUILD.2018.01.066.
42. P. H. Kuo and C. J. Huang, "A High Precision Artificial Neural Networks Model for Short-Term Energy Load Forecasting," *Energies*, vol. 11, no. 1, pp. 1–13, 2018, doi: 10.3390/en11010213.
43. T. Ahmad, H. Chen, and Y. Huang, "Short-Term Energy Prediction for District-Level Load Management Using Machine Learning Based Approaches," *Energy Procedia*, vol. 158, pp. 3331–3338, Feb. 2019, doi: 10.1016/J.EGYPRO.2019.01.967.
44. T. Ahmad *et al.*, "Supervised Based Machine Learning Models for Short, Medium and Long-Term Energy Prediction in Distinct Building Environment," *Energy*, vol. 158, pp. 17–32, Sep. 2018, doi: 10.1016/J.ENERGY.2018.05.169.

3 Exploiting the Flexibility Value of Virtual Power Plants through Market Participation in Smart Energy Communities

Georgios Skaltsis, Stylianos Zikos,
Elpiniki Makri, Christos Timplalexis, Dimosthenis
Ioannidis, and Dimitrios Tzovaras
Information Technologies Institute, Centre for Research
and Technology Hellas, Thessaloniki, Greece

CONTENTS

3.1	Introduction	56
3.2	Background on VPP Management, Objectives and Solving Approaches	58
	3.2.1 Dynamic VPPs' Concepts	58
	3.2.2 VPP Aggregation	59
	3.2.3 Objectives and Solving Approaches	60
3.3	Overall Concept – Virtual Power Plants and Integration With Markets	61
	3.3.1 Prosumers, DERs and the Aggregator	62
	3.3.2 Types of Markets	63
	3.3.3 DSO and the Traffic Light Concept	64
	3.3.4 Essential Technologies	65
3.4	VPP Formation Methodology	66
	3.4.1 Communication in Different Grid States	66
	3.4.1.1 Green State	67
	3.4.1.2 Yellow State	67
	3.4.1.3 Red State	67
	3.4.2 VPP Formation Algorithm	67
	3.4.2.1 VPP Formation in Green State	70
	3.4.2.2 VPP Formation in Yellow State	70
	3.4.2.3 VPP Formation in Red State	71
3.5	Conclusions	76
Acknowledgement		77
References		77

DOI: 10.1201/9781003257202-4

NOMENCLATURE

N	total number of assets
i	index of asset, $i = 1, \cdots, N$
t_{start}	time of activation
t_{end}	time of completion
t	time index, $t \in \{t_{start}, \ldots, t_{end}\}$
$b_i \in \{0,1\}$	binary decision variable
$F_{i,t}$	forecasted flexibility per asset and time slot
$F_{requested}$	requested amount of flexibility
$F_{requested, t}$	requested amount of flexibility per time slot
b_binary_i	corresponding binary variable of the heuristic algorithm
$b\_continuous_i \in [0,1]$	continuous decision variables of the heuristic algorithm
i_{min}	index for which $max(F_{i_{min}, t \in \{t_{start}, \ldots, t_{end}\}})$ and $b\_binary_{i_{min}} = 0$
b_final	final solution of the heuristic

3.1 INTRODUCTION

Increasing global energy demands, in combination with the need for CO_2 emission reduction, are rapidly changing the way that electrical energy is generated, distributed and consumed. The sustainability of modern smart energy systems is largely based on the penetration of distributed energy resources (DERs) to the grid. Thus, the efficient utilization and optimized operation of DERs within the grid are of high importance. Virtual power plants (VPPs) are bundles of DERs that create a sizeable capacity that becomes eligible to participate in wholesale (WS) power markets. The aggregator can provide various grid services such as frequency regulation and operating reserve capacity by optimizing a suitable portfolio of DERs. A VPP can include fast-response units such as supercapacitors and batteries, along with combined heat and power (CHP) and biogas power plants, and demand response (DR) resources to provide different flexibility services [1].

DERs are small- or medium-sized resources that are directly connected to the distribution network. Main categories of DERs are distributed generation, energy storage (e.g. stationary batteries) and demand side resources that include various types of controllable loads (Figure 3.1). Special attention is lately given to small-sized DERs that are located at residential prosumers' premises, and due to their large number, they have the potential to be used to form VPPs of considerably large energy capacity. VPP operators aggregate DERs to behave like a traditional power plant with standard attributes such as minimum and maximum capacity, ramp-up, and ramp-down and participate in markets to sell electricity or ancillary services (ASs). The VPP is controlled by a central information technology (IT) system where data related to weather forecasts, electricity prices in WS markets and the overall power supply and consumption trends are processed to optimize the operation of dispatchable DERs included in the VPP [2].

Energy management and control within a VPP can be performed in either centralized [3, 4] or decentralized manner [5]. In the former case, there is an approved central management entity that collects users' energy-related data and globally performs the optimizations, while in the latter case, energy management decisions take place

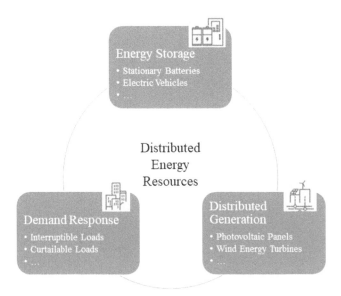

FIGURE 3.1 Distributed energy resources.

locally. The main advantages of the decentralized approach are less privacy issues, since there is no need to reveal detailed energy data to a centralized entity, as well as avoidance of severe single-point failures or bottlenecks. On the contrary, main disadvantages are the additional security and transparency requirements.

The aggregator is the ideal entity to create and manage VPPs, acting as a coordinator because of its role and objectives (Figure 3.2). Two important key enabling factors are (i) regulation and (ii) infrastructure and technologies [2]. As far as the regulation is concerned, a regulatory framework has to be established in order to allow aggregators to participate in the WS electricity market and the AS markets. Regarding infrastructure, smart meters and stable network connections are needed for real-time data acquisition from DERs. Moreover, support of two-way communication by the devices is necessary also for control and synchronization operations. Furthermore, load forecasting and forecasting power generation from renewable energy sources (RES) are both important for optimizing control of dispatchable DERs. Advanced forecasting methods are employed for determining these inputs.

VPPs offer a number of important advantages [6] and benefits for all involved stakeholders, as electrical energy is distributed in a more efficient and reliable way. First, they allow efficient management of energy demand and generation in a distributed environment, e.g. by enhancing self-consumption of participating prosumers. Furthermore, VPPs are valuable tools for the integration of RES towards their contribution to the balance of the grid. Another important advantage is that prosumers that are part of a VPP have safer and more cost-effective access to electricity markets as a group than they do individually, as the latter case makes meeting the minimum market entry limitations more challenging. Lastly, the integration of electric vehicle (EV) load management is another important aspect, as it combines the storage

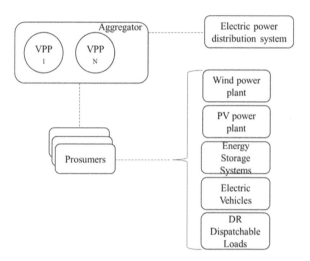

FIGURE 3.2 VPPs formulated and managed by the aggregator.

capacity and controllable loads that can be achieved by using the vehicle-to-grid (V2G) technology. All the aforementioned advantages contribute to the increasing interest in the development of VPPs worldwide. This becomes clearer from the fact that the global VPP market was valued at $1.3 billion in 2019 and is projected to reach $5.9 billion by 2027 [7].

This study presents the design concept, characteristics and technical requirements of a system running at the aggregator level for formulating VPPs dynamically in an automated way and exploiting aggregated small-scale DERs' flexibility in smart energy communities. Moreover, a methodology for dynamic VPP formation that takes into account the grid state according to the traffic light concept (TLC) is presented.

The rest of the study is organized as follows: Section 3.2 provides a background overview on topics related to VPP management, common problems to solve and common optimization approaches that are explored in the literature. Section 3.3 presents the integration of VPPs with markets, covering various trading scenarios towards better exploitation of the aggregated flexibility. Moreover, insights are given on requirements for implementation related to possible grid states and essential technologies. Section 3.4 provides the description of a VPP formation methodology, a heuristic optimization approach, as well as results from simulation experiments. Lastly, conclusions are summarized in Section 3.5.

3.2 BACKGROUND ON VPP MANAGEMENT, OBJECTIVES AND SOLVING APPROACHES

3.2.1 Dynamic VPPs' Concepts

Dynamic VPPs [8] is a new concept which integrates the dynamic aspects at all levels: Local level (i.e. per each RES), global level (i.e. interaction with elements of the grid and grid ASs), and economic aspect level (participation to electricity markets).

Exploiting the Flexibility Value of Virtual Power Plants 59

A dynamic VPP consists of a set of DERs along with a set of control operation procedures. The authors of [9] have presented an agent-based control method for dynamic VPPs self-adapting their unit set and operational plan to be able to trade products on a power market. The three coordination steps of aggregation, schedule optimization and continuous scheduling for dynamic VPP control are integrated into the behaviour of unit agents. In particular, for the active power delivery on day-ahead markets, the four following sequential phases take place: Dynamic VPP aggregation, market interaction, intra-dynamic VPP optimization and continuous scheduling. Dynamic VPP aggregation phase ensures that energy units are aggregated to VPPs in order to deliver active power products. In the market interaction phase, each dynamic VPP places its power products on the market and is informed upon acceptance of its offer. Intra-VPP optimization is performed considering the energy resources' operational status and predicted usage. Lastly, continuous scheduling ensures that rescheduling of the energy resources is performed if needed towards fulfilling the obligations related to the power contribution. As the potential of dynamic VPPs for power system control depends on the flexibility of each of the included assets, a generic representation of these flexibilities in well-defined data models is required. In this respect, the flex-offer representation [10, 11] describes energy flexibilities covering aspects such as start time and energy amount and facilitates the exchange of flexibility information between entities.

3.2.2 VPP Aggregation

VPP aggregation, or alternatively VPP formation, is the process of collecting and merging capacities of diverse dispatchable and non-dispatchable DERs, energy storage systems that may include EVs, controllable loads and DRs, in order to create a composite VPP, the capacity, characteristics and function of which are equivalent to a physical power plant [12]. The aggregation method is expected to ultimately impact on the performance and operation of the VPP; hence, various approaches have been proposed by different researchers. In [13], the authors proposed a virtual DER cluster (VDC) which constitutes an association of small-scale producers (usually DERs) that have agreed to operate on a common basis. A directory of DERs is built, from which dynamic VDCs can be formed. Cluster formation or aggregation is dynamic (occurring at hourly interval or day ahead) since energy demand and supply are dynamic processes.

Clustering methods have been applied to several cases within the VPP formation process. For instance, the authors of [14] examine clustering of distributed generators and consumers for optimal aggregation and management. The partition clustering algorithm k-means is applied to various scenarios in order to cluster resources for defining the remuneration of each identified group, which corresponds to a distinct DR program or remuneration group. In [15], the authors propose a dynamic distributed VPP formation method to cluster diverse DERs into heterogeneous VPP, based on their storage capacity and the owner's preference related to their willingness to provide power services. In that work, one of the VPPs provides the high-frequency power for regulating frequency and voltage. In addition, another VPP provides the required bulk power demand (low frequency). The proposed clustering algorithm allows offering a required energy capacity of the bulk VPP.

3.2.3 OBJECTIVES AND SOLVING APPROACHES

Regarding VPP management and scheduling of resources, interest in various objectives can be found in the literature. Two of the most common objectives are the maximization of the VPP's profit and the optimization of the bidding in markets. A bid includes information about the amount of power, the price, the region and the time interval during which a market participant is willing to buy or sell. Usually, constraints related to technical properties of DERs, market regulation requirements or grid limitations set by the distribution system operator (DSO) are considered in problem definition. Moreover, multiobjective optimization problems have been proposed to deal with different aspects concurrently. After the formulation of the problem and definition of the objective function(s) and the constraints, the determination of the optimization method to be used follows as a subsequent step. The goal of the optimization method is to return the optimal solution for the problem.

There are different types of optimization algorithms which can be categorized based on the nature of the variables (continuous or discrete) or based on the constraints (linear or nonlinear). Hereinafter, the derived main categories are linear programming (LP), mixed-integer linear programming (MILP), nonlinear programming and mixed-integer nonlinear programming. In LP, the objective functions as much as the set of the constraints on the decision variables are linear. Simplex method, a quite common LP algorithm, involves a linear function and several constraints are expressed as inequalities. This algorithm is generating and examining candidate vertex solutions and usually requires a negligible time to find the optimal one. Moreover, the column generation method is a technique which is utilized for solving MILP problems in case of a large number of variables in comparison with the number of constraints. This method is effective, considering the fact that it avoids enumerating all the possible elements like any traditional MILP algorithm. On the other hand, quadratic programming problems are a simple form of nonlinear problems, consisting of a convex or non-convex quadratic function of variables.

Many studies model the mathematical problem as a mixed-integer linear problem, through the fast execution time that it is provided. However, some authors formulate the VPP issue with nonlinear constraints, and due to nonlinearity, they apply several techniques trying to avoid the possibility of generating local optimal solutions. Regarding discrete optimization problems, a widely used algorithm is the branch-and-bound technique [16]. The branch-and-bound method recursively divides the search space into smaller spaces, called branches by using estimated bounds to limit the number of possible solutions. Nevertheless, the formed tree may become enormous; in that case, the available memory will be drained. Dynamic programming is another optimization method that breaks down a complex problem into a group of simpler sub-problems, solving them separately just once and then storing their solution. Therefore, the optimal solution will be a combination of sub-problems' outcomes.

Game theory constitutes a mathematical problem that models interactive decision-making processes and the decision-makers are called "players". Therefore, given that more than two players are participating, at each step, one player should take into consideration the actions of the other player(s) while constructing its own strategy.

The player is considered to be rational and has as a main focus on the maximization of its pay-off. Game theory is applied to solve problems in various fields such as economics and power systems [17].

Furthermore, the intention of heuristic methods is based on providing a solution (not an optimal one) in a reasonable time frame. Most of the widely common heuristic methods include trial and error, guesswork and the process of elimination, thus providing the opportunity of sufficiently solving the problem in the short term. In case of a large number of integer variables in addition to nonlinear constraints, this method is highly recommended, especially when the processing speed is equally important as the solution. The most widely used techniques which deliver the best outcome are particle swarm optimization (PSO) algorithm [18] and genetic algorithms (GAs) [19]. PSO algorithm is a stochastic optimization method based on swarm. Specifically, in order to find food, each member in the swarms is continuously changing the search pattern based on its own experience or other members' knowledge. Thus, PSO imitates the motion of birds' flocks, commences with a population of random solutions, moves in a high-dimensional space, and tries to discover the optimal solution. The advantages of PSO arise from the less calculation cost and less required memory. On the other hand, GA is based on the principles of biological evolution and selection. At each step, the GA randomly selects individuals to be parents and produces the children, the next generation. The algorithm utilizes three main types of rules at each iteration: Selection (the selection of the parents), the crossover (combining two parents) and mutation (the appliance of random changes to individual parents). GA also supports multiobjective optimization.

Generally, in multiobjective optimization problems [20], two or more optimization targets are conflicting, to wit the optimization of a single objective negatively affects another objective. Consequently, it is impossible to obtain one optimal solution which satisfies all the objective functions, hence a set of solutions (Pareto solutions) define the trade-off between the objectives. Nevertheless, sometimes multiobjective problems can be converted into single-objective optimization by methods such as weighted sum methods. In this method, the objective functions are summed up with the usage of weight coefficients, and then it is optimized. The weight given to an objective is assigned according to the objective's relative importance, after the normalization of the objective functions in order to maintain the same magnitude. Unfortunately, in case of a non-convex objective space, it is infeasible to preserve Pareto-optimal solutions. Another method concerns the ε-constrained method which restricts all the objectives to an epsilon value, while maintaining only one. This method is functional to both convex and non-convex problems.

3.3 OVERALL CONCEPT – VIRTUAL POWER PLANTS AND INTEGRATION WITH MARKETS

This section provides an introduction to the overall ecosystem where the aggregator plays a central role for aggregating prosumers' DERs forming dynamic VPPs within a local energy/flexibility market. Further details about the involved actors, including the distribution system operator, types of markets and essential technologies for implementation, are provided in this section.

In the review article [6], the authors provide an analysis of recent studies that propose VPP models by taking into account their interactions with different types of energy markets. The most important aspects that are considered are the formulation of the model, methods for solving mathematical problems, participation in different types of markets and the applicability to real case studies. Two of the main conclusions of that article were the following: (i) VPP modelling has become more complete by including more operating constraints; however, more advanced optimization methods are required to achieve an optimal solution due to increased complexity and (ii) distributed generation in the VPP resulted in more active participation in different types of markets (day-ahead, balancing markets, bilateral contracts).

3.3.1 Prosumers, DERs and the Aggregator

The prosumers that form an energy community, the aggregator and the DSO are the main involved actors that participate in the local market. The DERs of each prosumer are included in the aggregator's portfolio. In the general case, the prosumers are heterogeneous in terms of electricity consumption and behaviour patterns, for example, they can be involved as small-scale residential prosumers, office and commercial buildings or even small industries as prosumers. Common types of DERs, including those found in DR applications (Heating, ventilation, and air conditioning, domestic hot water, lights), are shown in Figure 3.3. Prosumers' smart devices interact with the aggregator via IT networking infrastructure supporting bidirectional connections that enable monitoring and profiling of their activity. A crucial dimension of a prosumer's activity is the so-called flexibility measure which is a means to quantify the availability of a certain prosumer to adjust its load. Naturally, flexibility is an abstract measure and it does not always represent the consumption or generation of a certain load.

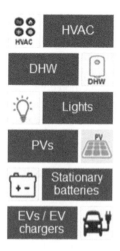

FIGURE 3.3 DERs, including those found in demand response applications.

Exploiting the Flexibility Value of Virtual Power Plants 63

The aggregator serves two vital roles in this framework. First, it interacts with prosumers' assets by receiving the load and flexibility estimation measurements and sending the proper control actions after the scheduling decisions have been made. The second role refers to the flexibility aggregation functionality which results in the optimized formation of VPPs and subsequent trading of their flexibility to the markets on behalf of the prosumers.

3.3.2 Types of Markets

Local energy market (LEM): An LEM is a concept that facilitates local energy trading between small-scale prosumers. Specifically, prosumers with a surplus of electricity production (e.g. from solar panels) are able to sell it to prosumers that are willing to buy it and increase their load. This procedure contributes to reduction of voltage fluctuations and relief of the distribution network. Although this concept can take place under the presence of a mediator, direct peer-to-peer trading is the current trend.

Local flexibility market (LFM): In the general case, LFM is a market-based tool for the DSO to control voltage violations and grid stress. In case the DSO forecasts a grid constraint violation, it reports a flexibility request to the operator of LFM (maybe the DSO itself or a third party), and in this way, an order is placed in this market. As a response to that order, aggregators are able to bundle flexibility offers from available prosumers and make a flexibility offer to the LFM. Prices are settled after the market is cleared; the flexibility is provided on behalf of the aggregators and the grid constraint violations are solved [21].

Regarding the implementation and operation of an LFM, two different approaches are analyzed in [22], namely the explicit and the implicit LFM. Explicit LFM resembles the characteristics of the default definition of LFM described previously. That is, the DSO aims to solve grid constraint violations by placing requests in the LFM, to which the aggregator is able to respond by bundling available prosumers' flexibility. In the implicit LFM case, there is no market platform for the LFM and activation of flexibility is achieved in the context of LEM, that is in the retail market. Specifically, it is implemented as a price-based control mechanism. That is, in contrast to the explicit case, the DSO does not publish any flexibility request in the platform. Instead, after having forecasted potential constraint violations and the location of the congestion points, it determines locationally differentiated grid prices. Naturally, the imposed prices reflect the grid constraints and the reaction of the prosumers is expected to indirectly solve the constraint violation problem.

ASs market: The ASs market facilitates the trading of services, and in general, the transmission system operator is the purchaser of products in this market, while sellers provide prequalified generation sources as well as DR and storage facilities in some cases. When ASs are considered mandatory support function, they are non-remunerated, but if this is not the case, the services can be paid through a regulated price, a pay-as-bid price or a common clearing price. In order to be able to make offers to ASs markets, methods need to be developed so that an aggregator is in position to evaluate the aggregated amount of service that can be delivered based on the type, size and location of the DERs included in the portfolio [23]. Some common categories of ASs are frequency control and voltage control services, system restoration after a fault in the grid and system control.

WS markets: These represent conventional WS electricity markets such as day-ahead market and intra-day market. The aggregator will have to decide upon which market to participate in, depending on the bidding horizon that it is willing to adjust to. Specifically, the bidding procedure in the day-ahead market occurs a day before delivery time and thus it should have sufficiently accurate forecasts about the prosumers' flexibility offers. On the other hand, intra-day market allows bidding in a much shorter period of time, and as a result, it is more suitable for trading flexibility which is by its nature challenging to forecast.

Power purchase agreements (PPAs): A PPA or bilateral contract constitutes a direct agreement for the sale of electricity with specified duration between two parties: An electricity power producer (the seller) and an electricity buyer. Both parties agree on a set of aspects such as the volume of power to be delivered, the minimum power that can be procured and the price. There are various forms of PPA as they are usually application specific and are established according to the needs of the buyer and the seller [24]. The main advantage of this type of agreement is price stability in the long term, as market varying prices are not employed. Use of PPAs can also be useful in certain cases where the DSO needs to procure flexibility that is not possible to be performed via the LFM, due to operational or time limitations (in an emergency).

3.3.3 DSO AND THE TRAFFIC LIGHT CONCEPT

The TLC has been adopted in many European smart grid discussions, which distinguishes between free market operation (green state), local market incentives to avoid network restrictions in advance (yellow state) and identified network restrictions (red state) [25–27]. Deployment of the TLC is necessary in order to ensure functional operation of the aggregator and clarify the prioritization of the offered services. As mentioned previously, there are three main operational states of the grid and they are described next.

In the green state, there are no constraint violations detected in the distribution grid by the DSO. As a result, LEM as well as AS/WS markets are active, whereas the LFM is paused. The operator of the LEM is responsible to clear and manage this market, and the actual role of the aggregator is to handle the assets that do not participate in LEM, optimally bundle their flexibility and trade them in AS/WS markets.

The yellow state is a temporary situation where constraints violations have been detected by the DSO. Depending on the type and urgency of those violations, the DSO decides to report a yellow state with either implicit or explicit LFM, which operates as described previously. In any case, in the context of the yellow state, LFM is operated alongside LEM, which means that peer-to-peer energy trading occurs simultaneously. Participation of the aggregator in AS/WS markets is paused. In practice, in the implicit LFM case, locationally differentiated prices are imposed by the DSO to all prosumers, who are also able to trade energy with each other in the context of LEM. On the other hand, when explicit LFM is implemented in the yellow state, prosumers are able to either participate in LEM or offer their flexibility to be exploited by the aggregator and thus be traded in the explicit LFM.

Exploiting the Flexibility Value of Virtual Power Plants 65

Finally, the red state implies that the grid is under stress due to violation of constraints, such as congestion and voltage violations. In such a case, the DSO is allowed to override market-based contracts and perform direct load control forcing loads to be switched off or reduced. This can be applied either directly via DSO's own infrastructure or by making use of services provided by the aggregator, as described in [22]. The latter case is discussed in more detail in the following sections.

In conclusion, grid state is a crucial parameter that drives the flexibility exchange and trading mechanisms that will be applied within the LFM as well as participation of the aggregator to the markets. More details about the behaviour of the aggregator and the VPP formation procedure in different states of the grid are presented in the following sections.

3.3.4 Essential Technologies

Various technologies and computation methods have to be involved and combined for the practical implementation of an automated VPP management system at the aggregator level that offers integration with different types of markets and is able to optimize operation and prosumers' profits. The four main ones are described next by highlighting their role: Internet of Things (IoT) and communication networks for asset monitoring and control, distributed ledger technology (DLT) for secure and transparent energy/flexibility transactions and market clearing, artificial intelligence (AI) algorithms for model training and flexibility forecasting and optimization and control algorithms.

IoT technology and communication networks: One of the key factors accelerating the transition from traditional energy systems towards smart grids is the integration of IoT devices. IoT equipment enables innovative ways to leverage devices, data and remote access, creating new business opportunities for various stakeholders. Monitoring of the grid is now performed in almost real time, providing the necessary data that are contributing towards the optimal control of the grid assets. A variety of communication protocols and networks are nowadays capable of dealing with large amounts of data, coming from multiple sources. Those data can be transmitted on high granularity and stored locally or on the cloud. After the data are processed by AI-based algorithms, the appropriate actuators are triggered, applying control actions to proper assets.

DLT: The centralized management of the electrical smart grids is turning out to be more and more inefficient over the years, highlighting the need for the development of decentralized solutions. Blockchain provides the technological infrastructure for the development of secure and reliable solutions in the energy sector. Several applications have already been implemented, focusing on energy trading, aiming at increasing the grid's flexibility and capability of adaptation to the energy demand, in a transparent and cost-effective manner. The operation of local flexibility and energy markets is also facilitated ensuring the execution of secure transactions for all the involved actors. Market clearing and automated settlement (rewards and penalties) are also achieved via the use of smart contracts.

Advanced forecasting methods: Load and generation forecasts derive the essential information that serves as an input for the management of the VPP operation.

More specifically, the forecasts reveal the flexibility potential that enables the optimal scheduling and facilitates energy exchange between the grid assets. Machine learning and deep learning techniques are utilized as state-of-the-art solutions for the implementation of the forecasting algorithms. Statistical methods usually serve as benchmarks for the more sophisticated AI-based methods. Unsupervised approaches also contribute using clustering techniques aiming to categorize asset usage profiles based on historical data. The creation of usage profiles may improve the predictive accuracy and, at the same time, reduce the computational cost. Another aspect that should be taken into consideration in flexibility estimation methods is the incorporation of external information that may affect the result. For example, the model should take into account that the occupants' comfort should not be violated and that probably creates stricter limitations to the amount of flexibility that can be delivered.

Optimization and control algorithms: Optimization techniques are used as the core feature for the optimal operation and control of the VPPs. Various techniques have been used in power systems optimization, including mixed-integer programming (both linear and nonlinear), dynamic optimization and GAs. An important part in the formulation of the optimization problem is the definition of its constraints. Those constraints can be either imposed by the physical equipment of the grid, or defined by the user in order to perform specific use cases. Optimal operation of the VPP can be perceived in multiple ways, depending on which stakeholder's benefit is attempting to serve. The most common ways of operation, typically defined by the problem's objective function, usually include the maximization of the economic benefit, the increase of self-consumption and the provision of maximum support to the grid through ASs.

3.4 VPP FORMATION METHODOLOGY

Based on the local grid network ecosystem and the formed local energy/flexibility markets that were described in the previous section, an approach for VPP formation by the aggregator is presented for facilitating efficient participation in different markets and services. As discussed in the previous paragraphs, efficient cooperation with the DSO is dictated via a traffic light scheme which is controlled by the DSO side, whereas the aggregator side is responsible for facilitating the requested services. In the following paragraphs, information that is necessary to be communicated with other components is discussed and the VPP formation methodology is deployed.

3.4.1 COMMUNICATION IN DIFFERENT GRID STATES

Different situations of the grid with respect to load and generation curves might require certain action from the side of the DSO. For example, stability issues are related to the condition of frequency and voltage across the network and, as an extension of that, with the power flow in different parts of the grid. It is part of DSO's responsibilities to take measures in order to robustify the network against such disturbances. Moreover, the aggregator is responsible to facilitate participation of end users to LFMs. Naturally, such procedures might require flow of necessary

Exploiting the Flexibility Value of Virtual Power Plants

information from the DSO. In the following paragraphs, the format of this communication is discussed so that the aforementioned services are provided efficiently.

3.4.1.1 Green State

According to the TLC, in green state, that is in normal operation conditions, the LEM and participation of the aggregator, on behalf of prosumers, in AS/WS markets are active. As described previously, support of LEM is not part of the aggregator's responsibilities and peer-to-peer flexibility trading takes place between prosumers under the surveillance of the market operator. However, participation in AS markets requires certain information depending on the type of AS that is provided, such as frequency control and voltage control or balancing. In ASs, the requested action is most commonly related to load reduction at a certain district of the grid and thus such information should be communicated. On the other hand, participation in the WS market, such as the intra-day market, does not require any information provided by the DSO. Instead, the aggregator should follow the rules and regulations of the marketplace of its interest in order to make bids for buying and selling energy.

3.4.1.2 Yellow State

According to the scheme proposed, in yellow state, that is when manageable problems are either detected or forecasted, the LFM is activated, either explicitly or implicitly. Additionally, LEM remains active, whereas participation of the aggregator in AS/WS markets is paused. Implicit LFM does not require any information that needs to be communicated through the aggregator, whereas in the explicit case, the aggregator communicates with the LFM platform in order to acquire information about flexibility requests of the DSO. The VPP formation strategy for each case is discussed in more detail in the following subsections.

3.4.1.3 Red State

This state is activated in cases of emergency. Specifically, when the grid, or part of it, is under stress, the DSO is able to perform direct control actions or request a specific action from the aggregator. In any case, bilateral contracts with the prosumers must be active, as the LFM is not engaged in this case. For example, it might request a load reduction of a certain amount of power in a particular district of the grid. The aggregator can handle such requests and facilitate congestion relief. An example of such a post is presented in JavaScript Object Notation (JSON) format in Figure 3.4.

As it can be seen, this event format includes all the information that is necessary to explicitly describe the desired action from the aggregator side. In the following section, we discuss the way that the aggregator exploits the incoming information in order to perform optimal VPP formation and decide on actions that need to be taken.

3.4.2 VPP Formation Algorithm

The logic behind forming VPPs is aligned with the initial purpose of the TLC, which is to optimally exploit available flexibility resources when the grid has stability margins and to be able to offer congestion relief in cases that the grid is under stress. Moreover, when no problem in the local grid is foreseen, selling excess

68   Virtual Power Plant Solution for Future Smart Energy Communities

```
{

            "id": "unique_id_key",
            "type": "GridNetworkStatus",
            "description": "A red status grid event",
            "dateTime": "2021-10-01T09:45:00Z",
            "sourceId": "DSO",
            "content": [
              {
                "forecastedStatus":
                  {
                    "regimeTF": "Red",
                    "refDateTime": "2021-10-01T11:00:00Z",
                    "location": "district_1"
                  },
                "currentStatus":
                  {
                    "regimeTF": "Red",
                    "flexibility_required": 5,            # in kW
                    "location": "district_1",
                    "delivery_horizon": 15,              # in minutes
                    "duration_of_DR": 60,                # in minutes
                    "start_time": "2021-10-01T10:00:00Z"
                    "end_time": "2021-10-01T11:00:00Z"
                  },
                "metadata": {}
              }
            ]
}
```

FIGURE 3.4   Information included in an indicative grid network status event.

flexibility of prosumers in markets for maximizing financial benefit is also desirable. A proposed methodology for addressing the dynamic VPP formation problem is illustrated in Figure 3.5. The methodology can be applied under any condition with regard to the grid state, providing that the inputs to be used in each case have been defined, as well as the objective function and optimization algorithm. The initiation of the process is either event-based when grid state changes or occurs periodically. At the first step, prosumer validation is performed in order to exclude the prosumers, and their assets/DERs, based on contractual information regarding their participation in the respective market or bilateral contracts. A similar procedure (DERs qualification) is followed in the next step for the DERs and any individual controllable assets, in order to exclude the ones that are not appropriate based on the three following aspects that are examined separately: Technical characteristics, availability and location. Technical characteristics refer to the type and control properties such as response times. Availability within the specified time interval is evaluated using the information about asset status and reservations for scheduled flexibility activations. Furthermore, the location in the grid is another important factor that is considered when it is available as an input parameter from the DSO for indicating the affected region. Location can be expressed in

Exploiting the Flexibility Value of Virtual Power Plants

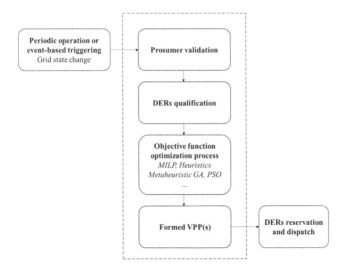

FIGURE 3.5 Steps of proposed dynamic VPP formation methodology.

geographical coordinates; however, location in the grid network is more relevant. The output of these two initial steps is a set of DERs and controllable assets that will be employed in the next step, namely the optimization process. At this point, the appropriate objective function is selected based on the objective(s), such as the maximization of profit or the aggregation of the requested amount of flexibility. After determining the DERs that will form the dynamic VPP, reservation and dispatch operations are applied.

Figure 3.6 presents the high-level inputs and outputs of the VPP formation algorithm. In the following subsections, details are provided on the actions that are performed in the different states of the grid, and a specific use case example is presented for the red grid state scenario that involves communication with the DSO in order to provide a given amount of flexibility.

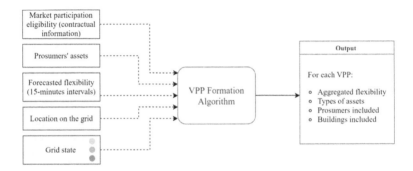

FIGURE 3.6 High-level inputs and outputs of the VPP formation algorithm.

3.4.2.1 VPP Formation in Green State

As described in the previous section, the grid being in a green state implies participation in multiple services. Specifically, participation of prosumers in LEM is projected, that is peer-to-peer flexibility trading, as well as participation in AS/WS markets. A prosumer is able to provide its consent if willing to be considered for participation in the AS/WS markets. Participation of prosumers in AS/WS markets is done via the aggregator.

The task of VPP formation in the green state is limited to optimally taking part in AS/WS markets. It might apply that in contracts for AS markets, such as the capacity market, an aggregator has to reserve in advance a certain amount of power that might be requested from a third party, e.g. the transmission system operator. In such a case, it is a priority to aggregate the required amount of power before trading any available flexibility in the WS markets. As an aggregation strategy for optimally gathering reserves for a potential AS markets request, the VPP formation algorithm proposed for the red state of the grid can be employed. More details about it can be found in Section 3.4.2.3.

Assuming that the aggregator has already formed a VPP for serving AS markets, there might be a surplus of flexibility offered by its prosumers. In the green state, this surplus can be traded in the WS markets. Among other alternatives, such as the day-ahead market, the intra-day market seems to be the most suitable given that the bidding horizon can be as short as 1 hour, which is very convenient for trading flexibility. In the literature, there are multiple approaches for optimal participation in the intra-day market [28], exploiting energy from renewables and battery storage. The problem is often tackled using multi-stage stochastic programming optimization, as in [29] and [30], which is a technique that employs the probability of occurrence for a number of generated probable scenarios in order to robustify the selected bidding strategy against uncertainties of forecasts. In the case that flexibility is the only resource to be traded in the WS market, profit maximization can be achieved by a plain greedy algorithm that trades all flexibility resources, or alternatively a more conservative approach that trades a percentage of it.

3.4.2.2 VPP Formation in Yellow State

As described in the section discussing the TLC, the yellow state of the grid might be applied either with implicit or explicit LFM. Moreover, LEM remains active, whereas participation of the aggregator in AS/WS markets is paused. In the implicit LFM case, there are no tasks that add responsibility to the aggregator. That is, the DSO is responsible to impose locationally differentiated prices to end users. In the meanwhile, end users are also able to take part in peer-to-peer energy trading in the context of LEM. On the other hand, yellow grid state with explicit LFM implies that the aggregator is responsible for bundling and trading available flexibility resources, that is flexibility of end users that do not take part in LEM. Explicit LFM is operated by the local flexibility market operator (LFMO) who provides the technical platform, where the DSO can trade flexibility with the aggregators. The LFMO is also responsible for the clearing and settlement procedure in this market. In this procedure, the aggregators announce their flexibility offers to the market operator

Exploiting the Flexibility Value of Virtual Power Plants 71

and the DSO makes certain flexibility requests. Then, it is the operator's responsibility to match offers with requests. There are a variety of methods for clearing similar LFMs, which are reviewed in [21].

Making offers as an aggregator in such a marketplace consists of bundling a certain amount of flexibility for the amount of time requested and setting a price for it. Given that other aggregators do similarly and act as market competitors, optimal bidding is not trivial. A conservative, yet straightforward, approach is to bundle a certain percentage of the available flexibility resources and set a price that is around the average electricity price in the corresponding WS markets. Naturally, whether a VPP formation strategy in explicit LFM is eventually successful or not strongly depends on the circumstances with respect to general demand and supply at a certain time. Moreover, the selected strategy can only be evaluated after the bidding procedure by comparing the selected prices with the prices of matched offers in a certain bidding procedure.

3.4.2.3 VPP Formation in Red State

In the case when the grid is in red state, the aggregator can be requested to offer a certain amount of flexibility for a certain amount of time in order to reduce the load of the grid. In the general case, the requested amount of flexibility might be described as a time series with different values in each 15-minute time interval. However, for sake of simplicity, we assume that the amount of flexibility that is requested is described by a single value and it is constant during the entire period that the grid is in the red state. The proposed algorithm can be easily extended to the general case.

Initially, the problem has to be formulated, that is to mathematically express the VPP formation task in case of red state of the grid, which is equivalent to finding the assets with a certain forecasted flexibility curve that, if aggregated and activated, optimally satisfy the red status flexibility request of the DSO. For the purpose of this problem formulation, the following integer linear programming (ILP) optimization scheme is proposed. Let us assume that there are N available assets and all satisfy the requirements of network district and flexibility delivery horizon of a certain post of the DSO. Additionally, each asset is accompanied by a forecast of its flexibility during the day denoted as $F_{i,t}$, where $i = 1, \ldots, N$ and t denote the time slot of the day. Optimizing the asset selection procedure is expressed by the following optimization scheme:

$$minimize \sum_{i=1}^{N} b_i \times F_{i,t},$$

$$s.t. \sum_{i=1}^{N} b_i \times F_{i,t} \geq F_{requested}, \; for \; t \in \{t_{start}, \ldots, t_{end}\}$$

The term $b_i \in \{0,1\}$ is a binary variable and denotes whether an asset is employed or not in order to serve the incoming DR. The term $F_{requested}$ denotes the requested amount of flexibility in kW. The timestamps t_{start} and t_{end} denote the start and end time of the DR and $\{t_{start}, ..., t_{end}\}$ is a set of discrete timestamps for which flexibility should be activated. The resulting VPP to which the task of serving the DSO flexibility request will be assigned consists of the assets for which $b_i = 1$. The proposed optimization scheme is discussed more in depth in the following paragraphs, employing a certain use case.

3.4.2.3.1 Use Case

In order to demonstrate the behaviour of this algorithm, we employ data that were synthetically created by a dataset describing the electric load of 64 consumers, provided by smart grid smart city (SGSC) project initiated by the Australian government [31]. In Figure 3.7, an example of the forecasted flexibility of a DER during 8 hours of a day is presented, with the time resolution equal to 15 minutes.

Moreover, we assume that the DSO posts a state update, and specifically the one described in Section 3.4.1.3. This post requests 5 kW of flexibility, with a duration of 60 minutes, starting at 10:00 in the morning. Such duration implies that, given the time resolution of the flexibility time series is 15 minutes, there are four discrete time slots for flexibility values within the 60-minute period. The proposed algorithm for these optimization parameters converges to the following solution depicted in Figure 3.8.

As it can be seen, 8 out of 64 assets, being represented in the graph by the coloured lines at the bottom, were selected to create the VPP that will serve the requested DR. The aggregated flexibility is, as expected, very close to the amount of requested flexibility, that is slightly higher than 5 kW and specifically equal to [5.01, 5.02, 5.01, 5.01] kW for the four quarters of that hour. It shall be noted that the aggregated flexibility is the sum of the individual flexibility values of each asset.

3.4.2.3.2 Computational Complexity

The approach presented earlier results in finding an optimal solution for the management of available resources, yet it comes with the cost of increased computational complexity. Specifically, ILP lies within the group of nondeterministic

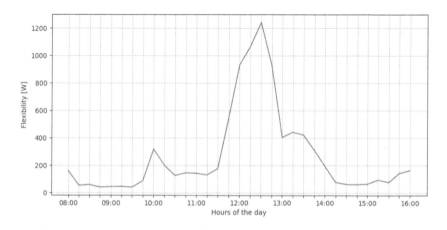

FIGURE 3.7 Example of 8-hour flexibility forecasting of a random asset.

Exploiting the Flexibility Value of Virtual Power Plants

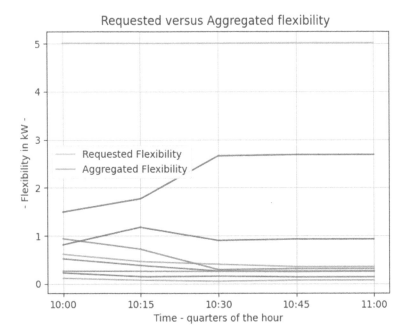

FIGURE 3.8 Aggregated flexibility after a DR request in red state.

polynomial-time complete problems, and for this particular optimization scheme, the time complexity is O(n!), which is equivalent to finding optimal solutions via exhaustive (brute-force) search. In practice, this implies that for a medium size problem, that is a relatively small number of available assets and a small set of feasible solutions, the time required for convergence grows to extreme values. For example, in the case described in the previous paragraph, increasing the set of feasible solutions by setting the amount of requested flexibility to 11 kW (instead of just 5 kW) would require around 4000 seconds for a commercial cpu to converge. It is noted that for a flexibility request of just 5 kW, the same cpu requires around 3 seconds, since the corresponding set of feasible solutions is much smaller. Given that in practice the number of available assets and the set of feasible solutions are expected to increase much further than the previous example, it is clear that computational complexity is an issue that needs to be addressed in order to make practical application possible.

ILP is a fundamental optimization problem and it is commonly met in applications such as telecommunication networks, production planning and scheduling problems. Owing to the fact that ILP comes at the cost of increased computational complexity, there has been an extensive effort to employ heuristic methods in order to reduce the time that is required for convergence to a solution. The authors in [32] describe the basic principles of the most representative heuristic methods for MILP optimization (a more general case of ILP). Specifically, they examine how different families of heuristic methods such as Folklore methods, OCTANE and line search methods, Pivoting methods and Feasibility pump seek to circumvent the inherent complexity of ILP problems by manipulating the optimization scheme. The final

74 Virtual Power Plant Solution for Future Smart Energy Communities

goal is to reduce the initial ILP problem to a convex problem, while ensuring that the final solution is good enough for the initial objective function. Next, a novel heuristic to reduce the computational complexity of the initial ILP problem is proposed. The algorithm is described step by step, which is as follows:

1. **Input:** Time series of available flexibility of each asset $F_{i,t}$ for $i = 1, ..., N$ and $t \in \{t_{start}, ..., t_{end}\}$. Moreover, $F_{requested, t}$ is given and it denotes the amount of requested flexibility for each time interval.
2. Initialize the parameters:

feasible_solution_not_found_yet ← *True,*
$\quad b\_binary_i \leftarrow 0, i = 1, ..., N.$

3. **while** *feasible_solution_not_found_yet* **do**:
4. Constraints $\leftarrow \sum_{i=1}^{N} b\_continuous_i \times F_{i,t} + b\_binary_i \times F_{i,t} \geq F_{requested, t}.$
5. Constraints $\leftarrow$ Constraints $+ \left[0 \leq b\_continuous_i \leq 1 \right], i = 1, ..., N.$
6. Constraints $\leftarrow$ Constraints $+ [b\_continuous_i = 0,$ for i that $b\_binary_i = 1]$
7. **Solve:** *minimize* $\sum_{i=1}^{N} b\_continuous_i \times F_{i,t},$ **s.t.** Constraints.

8. **for** $i = 1, ..., N$:

if $b\_binary_i = 0$:
$\quad b\_binary_i \leftarrow round(b\_continuous_i)$

9. **if** $\sum_{i=1}^{N} b\_binary_i \times F_{i,t} \geq F_{requested, t}$:
10. *feasible_solution_not_found_yet* ← *False*
11. *Employ b_binary which minimize the objective function.*
12. **If** *b_binary remains unchanged for two consecutive iterations:*
13. *Find i_{min} for which $b\_binary_{i_{min}} = 0$ and $max(F_{i_{min}, t \in \{t_{start}, ..., t_{end}\}})$ is the* minimum.
14. $b\_binary_{i_{min}} \leftarrow 1$
15. **end**
16. $b\_final \leftarrow b\_binary$

The proposed algorithm reduces the initial ILP problem to a number of iterative, plain linear programming problems. This is done by considering a continuous variable $b\_continuous_i$ for $i = 1, ..., N, b\_continuous_i \in [0,1]$, which is a convex set, as well as with convex constraints, and thus the overall iterative algorithm remains convex. The variable $b\_continuous_i$ is employed in order to represent the contribution of each asset which was initially described by the integer parameter $b\_binary_i \in \{0,1\}$. It should be highlighted that b_binary_i is dealt with as a parameter and not as an optimization variable in steps 4–7 and this is why convexity is preserved. At step 8 of the proposed algorithm, the parameters b_binary_i take values equal to the rounding of the converged $b\_continuous_i$ values. Conventional rounding assigns values in the interval

Exploiting the Flexibility Value of Virtual Power Plants 75

$[0,0.5]$ to 0 and values in the interval $[0.5,1]$ to 1. The algorithm can be easily modified at this step by considering a quantizer that splits the interval $[0,1]$ at a point different than 0.5. At step 9, the algorithm ensures that if the resulting binary solution is feasible, then it should stop searching. At step 11, the algorithm makes sure that the resulting final solution optimally satisfies the objective function, in case that the previously added assets led to an unnecessarily excessive amount of aggregated flexibility. In practice, this implies solving an additional ILP problem with size equal to the number of assets employed so far. Alternatively, in order to significantly reduce the size of this problem, the algorithm could only consider the assets that were most recently employed. Moreover, in the case when all the values of the converged continuous variables remain below 0.5, the algorithm prevents a potential infinity loop by employing the asset that its peak forecasted that flexibility value is minimum. This is performed in steps 13 and 14. It should be noted that the search algorithm performed in step 13 is theoretically an ILP problem and thus of high computational complexity. Nevertheless, the corresponding feasibility set is very small, since it only contains, at most, N elements, as many as the number of available assets, and as a result, the computational complexity is negligible. The final solution of the proposed method is equal to the resulting value of the b_binary parameter, which implies that assets with $b\_binary_i = 1$ will be employed for the VPP formation and assets with $b\_binary_i = 0$ will not.

It is clear that even though this method does not aim to find an optimal solution to the problem, it succeeds in finding a solution which is good enough while reducing the computational complexity. This can be better demonstrated through two examples, for which the results are presented in Tables 3.1 and 3.2.

TABLE 3.1

Performance of the Two Algorithms in a 5-kW Request Case

	Aggregated Flexibility (kW) per 15 minutes	Computation Time Required (seconds)
Exhaustive search	[5.01, 5.02, 5.01, 5.01]	2.31
Heuristic ILP	[5.356, 6.001, 5.117, 5.335]	0.01

TABLE 3.2

Performance of the Two Algorithms in a 12-kW Request Case

	Aggregated Flexibility (kW) per 15 minutes	Computation Time Required (seconds)
Exhaustive search	[12.01, 12.02, 12.01, 12.01]	Approx. 8000
Heuristic ILP	[13.433, 12.147, 12.861, 12.105]	0.01

76 Virtual Power Plant Solution for Future Smart Energy Communities

As it can be seen from the two tables, there is a critical difference in the behaviour of the two algorithms. As expected, in both DR signals, the exhaustive search achieves to aggregate an amount of flexibility, which is very close to the requested, since this algorithm converges to the optimal solution. However, optimality comes at the cost of an increase of the time that is required for convergence, which is clearer in the case that 12 kW is requested, that is in the case of a larger set of feasible solutions. Specifically, 8000 seconds is a non-practical amount of time for the application of VPP formation in the examined case, which needs to be served in near real time. On the other hand, the proposed algorithm lacks optimality, since it is obvious that there is a surplus of aggregated flexibility which reaches up to 20% of the requested amount in the case of the 5-kW DR signal. However, this surplus is expected to be significantly lower in cases where the requested amount of flexibility is larger, which is also the general case in practice. Most importantly, the amount of time that is required for the proposed algorithm to converge is very low, and specifically significantly lower than 1 second, as it can be seen in the two tables. In practice, the advantage of a fast response is crucial for performing VPP formation and thus the proposed algorithm seems to satisfy the need for a trade-off between optimality and responsiveness.

3.5 CONCLUSIONS

This work explored the characteristics and integration of VPPs in smart grid networks where participation in different types of markets is enabled within a local energy community. Various aspects and challenges on the exploitation of the flexibility offered by VPPs were highlighted, considering that it can be performed through both market participation and directly based on bilateral contracts. Moreover, related important topics that must be considered for the design and operation were highlighted, such as the operation of different types of markets, enabling technologies and optimization approaches.

Furthermore, a methodology for dynamic VPP formation towards optimal provision of aggregated flexibility has been presented, which is driven by the grid network status that is defined according to the TLC. The latter affects and defines the inputs that are considered by the dynamic VPP formation method. Focus has been given on VPP formation in intra-day flexibility delivery assuming that dynamic flexibility forecasting on asset or prosumer level is available. As far as the discovery of optimized solutions in feasible time is concerned, simulation experiments confirmed the applicability of a proposed heuristic method which was able to find solutions in a very short time for the tested use case. This is of high importance at emergency situations, as an exhaustive search would be impracticable even in cases where a relatively small number of DERs is engaged. To sum up, the automated VPP formation, settlement of performed transactions and flexibility trading can be implemented by combining technologies such as DLT, IoT networks, AI-based models and optimization methods, resulting in benefits for all involved participants.

ACKNOWLEDGEMENT

This work has been supported by the EU H2020 project PARITY, which has received funding from the European Union's Horizon 2020 Research and Innovation Programme under grant agreement number 864319.

REFERENCES

1. Ma, Z., Billanes, J. D., & Jørgensen, B. N. (2017). Aggregation Potentials for Buildings – Business Models of Demand Response and Virtual Power Plants. MDPI.
2. IRENA (2019). Innovation Landscape Brief: Aggregators. International Renewable Energy Agency, Abu Dhabi.
3. Kang, Y. (2017). Optimal energy management for virtual power plant with renewable generation. Energy and Power Engineering, 9(04), 308.
4. Naina, P. M., Rajamani, H. S., & Swarup, K. S. (2017, December). Modeling and simulation of virtual power plant in energy management system applications. In 2017 7th International Conference on Power Systems (ICPS) (pp. 392–397). IEEE.
5. Yang, Q., Wang, H., Wang, T., Zhang, S., Wu, X., & Wang, H. (2021). Blockchain-based decentralized energy management platform for residential distributed energy resources in a virtual power plant. *Applied Energy*, *294*, 117026.
6. Naval, N., & Yusta, J. M. (2021). Virtual power plant models and electricity markets – A review. *Renewable and Sustainable Energy Reviews*, *149*, 111393.
7. Virtual Power Plant Market by Technology and by End User: Global Opportunity Analysis and Industry Forecast, 2020-2027, May 2020, https://www.researchandmarkets.com/reports/5125498/virtual-power-plant-market-by-technology-and-by?utm_source=BW&utm_medium=PressRelease&utm_code=czcqjt.
8. Marinescu, B., Gomis-Bellmunt, O., Dörfler, F., Schulte, H., & Sigrist, L. (2021). Dynamic Virtual Power Plant: A New Concept for Grid Integration of Renewable Energy Sources. arXiv preprint arXiv:2108.00153.
9. Nieße, A., Beer, S., Bremer, J., Hinrichs, C., Lünsdorf, O., & Sonnenschein, M. (2014, September). Conjoint dynamic aggregation and scheduling methods for dynamic virtual power plants. In 2014 Federated Conference on Computer Science and Information Systems (pp. 1505–1514). IEEE.
10. Boehm, M., Dannecker, L., Doms, A., Dovgan, E., Filipič, B., Fischer, U., … & Tušar, T. (2012, March). Data management in the mirabel smart grid system. In Proceedings of the 2012 Joint EDBT/ICDT Workshops (pp. 95–102).
11. Valsomatzis, E., Hose, K., Pedersen, T. B., & Siksnys, L. (2015). Measuring and comparing energy flexibilities. In EDBT/ICDT Joint Conference (pp. 78–85). CEUR Workshop Proceedings.
12. Adu-Kankam, K. O., & Camarinha-Matos, L. M. (2018). Towards collaborative virtual power plants: Trends and convergence. *Sustainable Energy, Grids and Networks*, *16*, 217–230.
13. Botsis, V., Doulamis, N., Doulamis, A., Makris, P., & Varvarigos, E. (2015, August). Efficient clustering of DERs in a virtual association for profit optimization. In 2015 Euromicro Conference on Digital System Design (pp. 494–501). IEEE.
14. Spínola, J., Faria, P., & Vale, Z. (2015, June). Scheduling and aggregation of distributed generators and consumers participating in demand response programs. In 2015 IEEE Eindhoven PowerTech (pp. 1–6). IEEE.
15. Zhang, R., & Hredzak, B. (2020). Distributed dynamic clustering algorithm for formation of heterogeneous virtual power plants based on power requirements. *IEEE Transactions on Smart Grid*, *12*(1), 192–204.

16. Di Somma, M., Graditi, G., & Siano, P. (2018). Optimal bidding strategy for a DER aggregator in the day-ahead market in the presence of demand flexibility. *IEEE Transactions on Industrial Electronics, 66*(2), 1509–1519.

17. Wang, Y., Ai, X., Tan, Z., Yan, L., & Liu, S. (2015). Interactive dispatch modes and bidding strategy of multiple virtual power plants based on demand response and game theory. *IEEE Transactions on Smart Grid, 7*(1), 510–519.

18. Tao, L., Liu, J., An, Q., Zhang, T., & Wang, J. (2020, July). Aggregating energy storage in virtual power plant and its application in unit commitment. In 2020 IEEE/IAS Industrial and Commercial Power System Asia (I&CPS Asia) (pp. 768–773). IEEE.

19. Wang, J., Yang, W., Cheng, H., Huang, L., & Gao, Y. (2017). The optimal configuration scheme of the virtual power plant considering benefits and risks of investors. *Energies, 10*(7), 968.

20. Hadayeghparast, S., Farsangi, A. S., & Shayanfar, H. (2019). Day-ahead stochastic multi-objective economic/emission operational scheduling of a large scale virtual power plant. *Energy, 172*, 630–646.

21. Jin, X., Wu, Q., & Jia, H. (2020). Local flexibility markets: Literature review on concepts, models and clearing methods. *Applied Energy, 261*, 114387.

22. Pressmair, G., Kapassa, E., Casado-Mansilla, D., Borges, C. E., & Themistocleous, M. (2021). Overcoming barriers for the adoption of local energy and flexibility markets: A user-centric and hybrid model. *Journal of Cleaner Production, 317*, 128323.

23. Oureilidis, K., Malamaki, K. N., Gallos, K., Tsitsimelis, A., Dikaiakos, C., Gkavanoudis, S., ... & Demoulias, C. (2020). Ancillary services market design in distribution networks: Review and identification of barriers. *Energies, 13*(4), 917.

24. Shabanzadeh, M., Sheikh-El-Eslami, M. K., & Haghifam, M. R. (2015). Decision making tool for virtual power plants considering midterm bilateral contracts. In Iranian Regulation CIRED Conf. and Exhibition on Electricity Distribution (pp. 1–6).

25. Deutsch, T., Kupzog, F., Einfalt, A., Ghaemi, S., & Austria, A. A. A. (2014, June). Avoiding grid congestions with traffic light approach and the flexibility operator. In CIRED Workshop (No. 0331, pp. 11–12).

26. BDEW – German Association of Energy and Water Industries: Smart Grid Traffic Light Concept – Design of the Amber Phase. https://www.bdew.de/media/documents/Stn_20150310_Smart-Grids-Traffic-Light-Concept_english.pdf (2015), accessed online 20 October 2021.

27. Pack, S., Kotthaus, K., Hermanns, J., Paulat, F., Meese, J., Zdrallek, M., & Braje, T. (2018, September). Integration of smart grid control strategies in the green phase of the distribution grid traffic light concept. In NEIS 2018; Conference on Sustainable Energy Supply and Energy Storage Systems (pp. 1–5). VDE.

28. Nguyen, H. T., Le, L. B., & Wang, Z. (2018). A bidding strategy for virtual power plants with the intraday demand response exchange market using the stochastic programming. *IEEE Transactions on Industry Applications, 54*(4), 3044–3055.

29. Heredia, F. J., Cuadrado, M. D., & Corchero, C. (2018). On optimal participation in the electricity markets of wind power plants with battery energy storage systems. *Computers & Operations Research, 96*, 316–329.

30. Crespo-Vazquez, J. L., Carrillo, C., Diaz-Dorado, E., Martinez-Lorenzo, J. A., & Noor-E-Alam, M. (2018). A machine learning based stochastic optimization framework for a wind and storage power plant participating in energy pool market. *Applied Energy, 232*, 341–357.

31. Smart Grid, Smart City Customer Trial Data [Online]. Available: https://data.gov.au/data/dataset/4e21dea3-9b87-4610-94c7-15a8a77907ef.

32. Fischetti, M., & Lodi, A. (2010). Heuristics in mixed integer programming. In Wiley Encyclopedia of Operations Research and Management Science. John Wiley & Sons, Inc.

4 Renewable Energy Community VPP Concept Design and Modelling for Sustainable Islands

R. Garner, G. Jansen, and Z. Dehouche
Department of Mechanical and Aerospace Engineering,
Brunel University London, Uxbridge, United Kingdom

CONTENTS

4.1 Introduction 80
    4.1.1 The Virtual Power Plant Concept 81
    4.1.2 Modelling Techniques and Optimization 82
    4.1.3 Applications on Geographical Islands: Formentera, Spain 82
    4.1.4 Hybrid Energy Storage System Sizing and Novelty 82
    4.1.5 Objectives 84
4.2 Methodology 84
    4.2.1 System Design 84
    4.2.2 Systems Modelling 85
        4.2.2.1 Photovoltaic Solar Model 85
        4.2.2.2 Lithium-Ion Battery 87
        4.2.2.3 Hydrogen Electrolyser and Fuel Cell 88
    4.2.3 Logic Control System 90
    4.2.4 Economic and Environmental Impacts 91
4.3 Results 94
    4.3.1 Economic Optimization 94
    4.3.2 End-User Energy Costs 97
    4.3.3 Environmental Impact Assessment 97
    4.3.4 VPP Analysis 98
4.4 Conclusion 99
Acknowledgement 100
References 100

DOI: 10.1201/9781003257202-5

4.1 INTRODUCTION

The market for distributed renewable energy systems (RES) has been increasing considerably in recent decades due to several economic, commercial, and climate-related factors [1]. The 2021 emissions gap report from the UN environment programme concluded that nations have not acted fast enough to limit global temperature rises below 1.5°C to comply with the objectives of the 2015 Paris Agreement [2]. Under more realistic scenarios, the global average temperature would rise far above the 2°C aspirational limit towards 2.7°C, causing widespread devastation [3], as depicted in historical surface temperature rise in Figure 4.1. Several complied results in the IPCC 2021 report indicate that a global overhaul of the way in which nations generate and consume energy is urgently needed [4]. To achieve any notable effect on the continually increasing temperatures, ambitious and drastic changes need to be made to the way in which the current model of electricity production operates. The energy supply in the UK produced 104.2 M tonnes of CO_2 equivalent in 2018, accounting for 23.2% of the total greenhouse gas emissions, only being surpassed by the transport industry with 27.6% share [5]. Assuming that the emissions share values are similar for other European countries, there is scope for major improvements in the emissions reduction from grid energy production.

Many countries have already taken strides in introducing RES into the national grid energy network. For example, in the UK renewable energies increased their share of power generation to 37% as of 2019, as seen in Figure 4.2, reducing the requirement for oil and coal power [6]. However, the energy balance will become less efficient over time and require significant investment in grid infrastructure [7] and is still nowhere near enough to reduce global climate effects. Energy storage systems (ESSs) must be utilized to smooth out peak and valleys in the supply and demand of energy [8]. This work sets out to analyse how the implementation of a

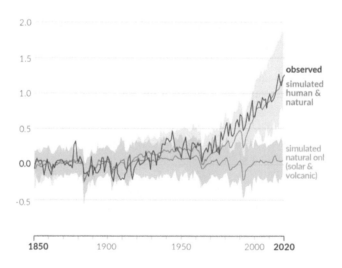

FIGURE 4.1 Change in global average surface temperature compared to a simulated natural only scenario [4].

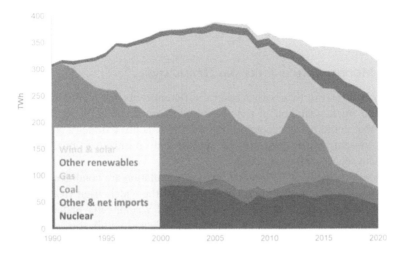

FIGURE 4.2 UK electrical supply by fuel type, 1990–2020 [5].

virtual power plant can not only present significant reductions in the environmental impact of electrical power consumption but also provide a cost-effective alternative to the current energy delivery approach.

4.1.1 THE VIRTUAL POWER PLANT CONCEPT

A VPP essentially combines the power outputs of several heterogeneous distributed energy resources (DER) sources along with ESS to function as a single energy source, usually consisting of both dispatchable and non-dispatchable sources [9]. Non-dispatchable sources including PV solar and wind turbines form the base power generation, then an ESS can be used to store the excess energy and release at a later stage. This in effect reverses the top-down approach of the current energy market model and puts the power generation and operation into the hands of local energy generators and consumers.

Early VPP concepts presented a variety of different methods for achieving the targeted flexibility of services and increased usage of distributed RES. The pilot project FENIX VPP's objective was to demonstrate both the technical and commercial architectures required to increase widespread adoption DER in Europe as a solution for a green energy future [10]. The aim in alignment with the simple definition of a VPP is to represent DER as a single system which both generates and consumes energy. The Edison VPP takes the concept further to incorporate flexible EV charging facilities to reliably balance the inconsistencies between supply and demand [11].

Several other VPP pilot studies and commercial project have set out to use small-scale PV solar installations within residential and community settings to varying degrees of success. These systems include the Con Edison VPP [12], Sonnen Community [13], and most recently the successfully commercialized AGL VPP in South Australia [14]. One of the key components to success is in the inclusion of community level and home-installed energy generation and storage systems which can facilitate the natural expansion of system capacity over time and provide tangible

82 Virtual Power Plant Solution for Future Smart Energy Communities

benefits for the consumer. This is one of the main objectives to be presented within the economic assessment of the VPP designed within this work.

4.1.2 Modelling Techniques and Optimization

As described in [15], there exists several different tools and methodologies for the design and implementation of time-based and parametric renewable energy components that could be incorporated into a VPP. A control strategy or energy management system (EMS) must also be applied and optimized to produce the most appropriate scenarios for validation.

The main methods of VPP modelling and simulation are conducted in commercially available RES modelling software, such as the widely used HOMER Energy. Examples of the usage of HOMER for modelling PV solar cells, battery, and hydrogen systems exist in [16, 17]. The modelling conducted within this work is instead conducted within the MATLAB/Simulink environmental due to the additional capabilities in model dynamics and the parameterization of individual component characteristics. Further examples of the use of this software for VPP and renewable system design exist in [18, 19].

4.1.3 Applications on Geographical Islands: Formentera, Spain

One of the biggest challenges in the decarbonization of European countries is in the development of renewable energies on geographical islands, which often rely on low-capacity sea cables to receive energy from the mainland or use on site diesel generators for additional power in remote areas. Island states such as Malta and Cyprus have reported the highest energy and network costs [20] in Europe, likely due to the requirement for mainland energy trade and fuel imports. Considering its landmass, Cyprus also has one for the highest greenhouse gas emissions per capita, which is approximately 31% over the European average [21]. Geographical islands also often lack the necessary facilities and expertise to develop potential smart grid concepts to increase sustainability and reduce reliance on imported energy.

To understand the macroeconomic and environmental effects on an individual island basis, a case study was to be performed. Formentera is a small semi-autonomous region within the Balearic Islands of Spain which is located south of Ibiza and of the Iberian Peninsula of Portugal and Spain, as displayed in Figure 4.3. The island has a temperature Mediterranean climate, as shown in the average monthly temperature data and solar irradiance levels shown in Figure 4.4. It is known that the island itself relies heavily on diesel and gas generators to provide additional energy supply during the summer season, which adds to the overall energy cost and significantly increases the emissions impact of electrical consumption, as displayed in Table 4.1.

4.1.4 Hybrid Energy Storage System Sizing and Novelty

A key requirement to increasing the flexibility of the VPP system is to include an ESS to store excess energy generation which would otherwise apply additional stress to the grid, and then release the energy when required. The latter moments

Renewable Energy Community VPP Concept Design and Modelling

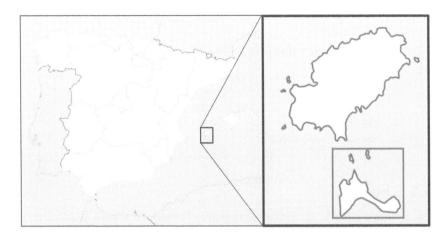

FIGURE 4.3 Iberian Peninsula with indicated location of Formentera (right).

are most likely when renewable energy is no longer available in excess, increasing the self-consumption of connected buildings. Batteries are often described as the most appropriate energy storage device for use in a VPP, as the increasing demand for the technology in personal electronics and EVs has driven down costs considerably. Even given this consideration, the utilization of batteries for large standalone grid ESSs can be prohibitively expensive [8], due to maintenance costs and lifespan issues. For this reason, a regenerative hydrogen system was designed for use in parallel to the battery to alleviate some of the identified shortcomings. Hydrogen system benefits from a higher energy density and lower self-discharge losses. The chosen methodology builds upon the works of [15, 19, 22] centred around the introduction of a hybrid hydrogen and battery ESS for VPP implementation by designing a novel state-based control logic, and an optimization method to maximize the economic benefits of the regenerative system. Given these specific aims, the objectives can be laid out as follows.

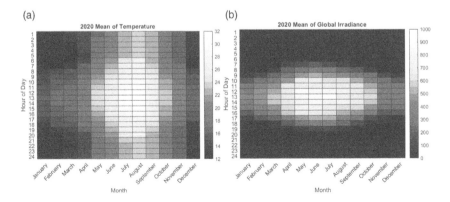

FIGURE 4.4 (a) Hourly mean temperature (°C) variation for each month and (b) irradiance (W/m^2) variation.

TABLE 4.1
Power Generation Mix on the Island

Generator	Capacity (MW)	Usage
PV solar farm	2	All time
Diesel generators	12 × 1.5 (18 total)	All time
Gas-oil turbine	13	Peak season only
Sea cable interconnection	NA	All time

4.1.5 OBJECTIVES

- Design a suitable small-scale community-based VPP system utilizing real production and consumption data from the field
- Implement a hybrid ESS combining the outputs of a lithium battery and regenerative hydrogen system to balancing the output of the combined PV solar system and community building loads
- Optimize system sizing to maximize the economic benefits, including the levelized cost of electricity (LCOE) and demonstrate reductions in the environmental impact (emissions intensity)

4.2 METHODOLOGY

4.2.1 SYSTEM DESIGN

The VPP system was designed and modelled to validate the performance of the individual components in relation to the described performance objectives. The modelling and simulation were conducted with the MATLAB/Simulink environment using component constructed from first principles and electrical algorithms contained within the specialized power systems (SPS) extension.

The model consists of four locally situated community buildings each fitted with PV solar system, capable of both self-consumption and sending excess power production to the grid. These four buildings include a council office, community centre, school, and a local football field. The hybrid ESS would then be connected and located near to the community buildings to serve as an energy buffer by storing excess energy and releasing when required. The locations of the real community buildings are shown in Figure 4.5, with the capacities of PV solar. The total capacity has been artificially increased slightly to allow for additional flexibility in the excess generation that would be made available. The building dimensions and available installation space was considered using the Helioscope PV solar planning software such that the increased capacity is realistic to the true environment. The irradiance data collected also shows a relatively high potential for PV solar, so increasing this further would be the most cost-effective method of reducing reliance on the grid connection.

The optimal storage sizing and system control logic is based on a number of key community VPP characteristics, including most importantly the PV solar generation

Renewable Energy Community VPP Concept Design and Modelling 85

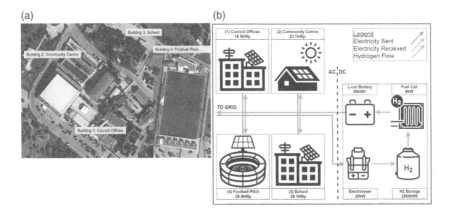

FIGURE 4.5 (a) Energy Community buildings with increased PV solar installation and (b) VPP schematic with system sizes.

available and the building load profiles. The samples of the real building loads were used in combination with consumption characteristics of similar usage buildings within the United Kingdom to construct 12-month load profiles for each of the four buildings. The graph in Figure 4.6 displays the building load profiles and PV solar generation for a one-week sample.

4.2.2 Systems Modelling

4.2.2.1 Photovoltaic Solar Model

Harnessing the power of the sun has for a long time been one of the fastest growth RES among the various technologies [23] and is set to increase with the further cost reductions in manufacturing and installation. The mathematical accuracy and modelling capability is therefore constantly updating as research find new methods for representing the characteristics of solar cells. As such, it is well known that a PV module can be represented as a simplified electrical circuit containing a photodiode as a variable current source within a single diode circuit, as shown in Figure 4.7. The most up-to-date equations for calculating each parameter at time of writing

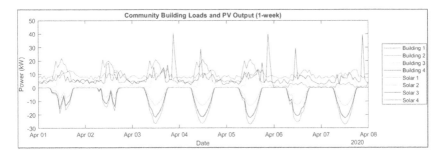

FIGURE 4.6 Building loads and PV solar generation one-week sample.

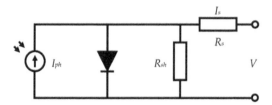

FIGURE 4.7 Solar cell equivalent circuit diagram.

were available from [24], and the model is an extension of the work completed by Motahhir et al. and presented in [25].

The photocurrent I_{ph} is found as a function of the short-circuit current I_{sc}, short-circuit temperature coefficient k_i, cell temperature T_p, cell reference temperature T_{ref}, and total solar irradiance G.

$$I_{ph} = \frac{G\left[I_{sc} + k_i\left(T_p - T_{ref}\right)\right]}{1000}$$

The diode saturation current I_0 is then a function of the reverse-saturation current I_{rs}, cell temperature and reference temperature, electron charge constant q, semiconductor band gap energy E_{g0}, diode ideality factor n, and the Boltzmann constant K_b.

$$I_0 = I_{rs}\left(\frac{T_p}{T_{ref}}\right)^3 \exp\left[\frac{qE_{g0}\left(\frac{1}{T_{ref}} - \frac{1}{T_p}\right)}{nK_b}\right]$$

The reverse-saturation current I_{rs} is found as a function of the short-circuit current, electron charge constant, open-circuit voltage V_{oc}, diode ideality factor, number of cells in series N_s, Boltzmann constant, and the cell temperature, described as the following:

$$I_{rs} = \frac{I_{sc}}{\exp\left[\frac{qV_{oc}}{nN_s K_b T_p}\right] - 1}$$

The current passing through the shunt resistor is determined using the fundamental Kirchoff's law, with the load voltage V, load current I, series resistance R_s, and shunt resistance R_{sh}. The equation is given below:

$$I_{sh} = \left(\frac{V + IR_s}{R_{sh}}\right)$$

The output current is given as a relationship between the photo current, saturation current, electron charge constant, load voltage, output current, series resistance,

Renewable Energy Community VPP Concept Design and Modelling

diode factor, Boltzmann constant, number of modules in series, cell temperature, and shunt current.

$$I = I_{ph} - I_0 \left[\exp \left(\frac{q(V + IR_s)}{nK_b N_s T_p} \right) - 1 \right] - I_{sh}$$

4.2.2.2 Lithium-Ion Battery

The modelling of ESSs is crucial in understanding the flexibility requirements and availability of the VPP to provide additional services, including temporal storage and smoothing of RES power output. Many different battery chemistries are potential suitable for the VPP application, but most cited is the lithium-ion battery. This is due to their already prevalent popularity in consumer electronics, electric vehicles (EV), and uninterruptible power supplies (UPS), with a high energy-to-weight ratio, high open-circuit voltage, low self-discharge, and no memory effect [26].

The model described in this section forms the foundation for the component used in the MATLAB/Simulink software, the equations of which are explained in detail in [27]. The model is based on an equivalent electrical circuit, similar to the PV panel methodology, of which the most important components are an ideal voltage source and a specified internal resistance [28]. A modified version of the Shepard battery model [29] is integrated into this design in order to both describe charge and discharge models based on the terminal voltage, open-circuit voltage, internal resistance, discharge current, and state-of-charge. For lithium-ion batteries, the equations for charge and discharge are given as the following:

Discharge:

$$V_{batt} = E_0 - R \cdot i - K \frac{Q}{Q - it} \cdot \left(it + i^* \right) + A exp(-B \cdot it)$$

Charge:

$$V_{batt} = E_0 + R \cdot i - K \frac{Q}{it - 0.1 \cdot Q} \cdot i^* + K \frac{Q}{Q - it} \cdot it - A exp(-B \cdot it)$$

where

V_{batt} is the non-linear voltage (V)
E_0 is the constant voltage (V)
K is the polarization resistance or polarization constant (Ω)
i^* is the low-frequency current dynamics (A)
i is the cell current (A)
it is the extracted capacity (Ah)
Q is the total capacity (Ah)
A is the exponential voltage (V)
B is the exponential capacity (A/h)

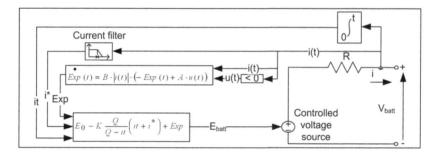

FIGURE 4.8 Lithium-ion battery equivalent circuit diagram under charging condition [27].

The formulae are arranged in the model as shown in Figure 4.8. It should be noted that lithium-ion batteries also do not need to consider the non-linear exponential function that factors in hysteresis effects in charge and discharge, as is for other battery chemistries.

This simple model does include several assumptions, such as constant internal resistance, constant capacity, and no temperature, self-discharge or ageing effects. The latter assumptions can be considered, however the physical phenomena that describe these effects are complex and largely outside of the scope of the work.

4.2.2.3 Hydrogen Electrolyser and Fuel Cell

The hydrogen system considered for modelling is the Proton Exchange Membrane Fuel Cell (PEMFC) and electrolyser system, both of which operate using the same fundamental physical processes based on Faraday's law of electrolysis [19]. The principal chemical redox reactions of converting water into hydrogen are shown below:

$$2H_2O \rightarrow 4H^+ + O_2 + 4e^- \text{ (Anode)}$$

$$4H^+ + 4e^- \rightarrow 2H_2 \text{ (Cathode)}$$

During operation, the electrolyser experiences three main physical losses that dictate the total stack voltage V_{el}, which increases as the efficiency decreases:

$$V_{el} = E_{nerst} + E_{act} + E_{ohm} + E_{conc}$$

The activation losses describe the potential that is lost due to the initial activation energy required for the redox reaction to occur.

$$E_{act} = -\left(\frac{RT}{\alpha F}\right)\ln(i_0) + \left(\frac{RT}{\alpha F}\right)\ln(i)$$

Ohmic potential losses occur naturally due to the linear resistive losses of the system.

$$E_{ohm} = iR_i$$

When the electrolyser runs at a high load, higher than normal current densities can lead to concentration losses, where the transportation flow of reactants cannot occur fast enough for the reaction process.

$$E_{conc} = \frac{RT}{nF} \ln\left(\frac{i_l}{i_l - i}\right)$$

The hydrogen production rate can then be calculated with the following:

$$f_{H_2} = N_p \eta_f \left(\frac{I}{nF}\right) \text{ where } \eta_f = 96.5 \exp\left(\frac{0.09}{i} - \frac{75.5}{i^2}\right)$$

The PEMFC can be considered the opposite of the electrolyser as the reactants and products are reversed, so the potential losses equation is given as the following:

$$V_{el} = E_{nerst} - E_{act} - E_{ohm} - E_{conc}$$

And finally, the hydrogen consumption flow is found as the following:

$$f_{H_2} = S\left(\frac{IN_p}{nF}\right)$$

Once these equations and components are reacted within the software environment, the four-building VPP system can be displayed graphically, as shown in Figure 4.9. On the left side are the four building loads and individual PV solar systems represented,

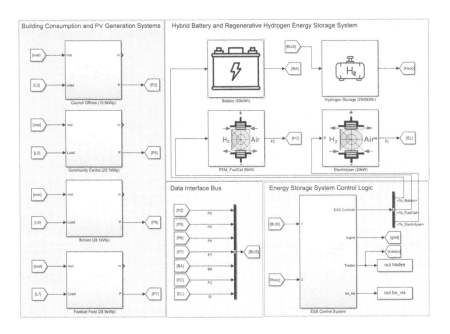

FIGURE 4.9 VPP system model represented in MATLAB/Simulink.

and on the right in descending order, displayed are the lithium-ion battery, PEM fuel cell, PEM electrolyser, and hydrogen storage system.

4.2.3 Logic Control System

The VPP control system is fundamentally driven by the measurement of excess energy produced by the combined PV solar output of all buildings during full self-consumption. This value can be described below:

$$P_{excess} = \sum P_{PV_n} - \sum P_{con_n}$$

where P_{excess} is the calculated excess power available, P_{PV_n} is the PV power output, and P_{con_n} the power consumption, both for a given building n.

For example, if the system detects that energy is available in excess of demand, the control logic will set the states of the battery, electrolyser based on a hierarchical state-based flow, where battery charging is prioritized, followed by the activation of the electrolyser once the maximum charge percentage is reached. This logic is shown in the diagram below:

When $P_{excess} > 0$:

If the excess power is not available, the controller will again measure the battery state of charge, and discharge of the SOC is above the depth of discharge limit. Once at this limit, the fuel cell is activated to charge the battery, which is then resupplying power to the load. Any additional power required is taken from the grid connection in this instance.

When $P_{excess} \leq 0$:

It can be noted that using this method, the fuel cell is never used to directly follow the load. This is because the efficiency of the fuel cell can vary greatly depending on the size of the load drawn. Operating the fuel cell at higher or lower loads than the rated output can limit its lifespan and damage the internal materials. For this reason, the fuel cell is set to its most efficiency power output and used to charge the battery, as the battery can more easily follow the load requirements of the buildings without large differences in efficiency. The diagram below displays the complete control flow logic for the hybrid ESS, including the control outputs for the battery, electrolyser, and fuel cell.

There is an extra transient control loop as shown in Figure 4.10 which is used to limit the cycling of battery depth of discharge under the influence of the fuel cell to 50% to increase lifespan and limit cycling effects. Theoretically this parameter

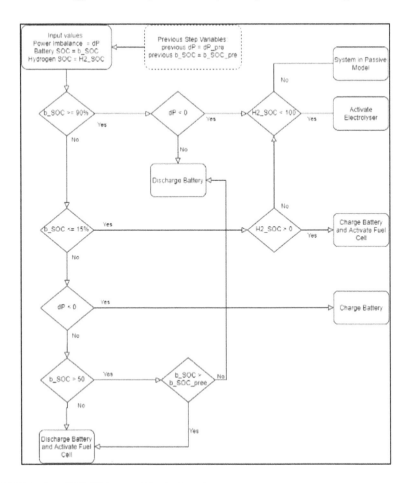

FIGURE 4.10 VPP ESS control logic flow chart.

could then be tuned to maximize the usefulness of the battery whilst also minimizing the degradation of performance. Figure 4.11 displays the control logic as a state flow chart.

4.2.4 ECONOMIC AND ENVIRONMENTAL IMPACTS

One of the key objectives of this work was to optimize the performance of the hybrid battery-PEMFC system to minimize the impacts of energy cost and emissions for the consumer.

For the economic assessment, the data first to be gathered was the energy cost currently paid by the consumers for their grid energy. This will be used as a benchmark for any improvements made with the implementation of the VPP. As the cost of energy can vary greatly depending on the factors such as the whole sale cost, utility network, and chosen tariff, the purposes of this work a flat-rate cost of 30 €cents/kWh

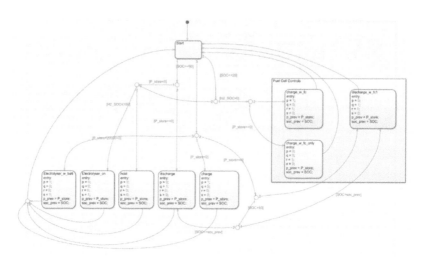

FIGURE 4.11 VPP ESS state flow chart implementation.

was used, in line with the fluctuating costs of electricity in Spain, Germany, and the United Kingdom.

The comparison was made to the LCOE of the VPP system, which requires an economic assessment of each component and its lifetime-predicted usage. Calculations of LCOE often depend on estimations of system lifespan and usage, often referred to as the capacity factor. The capacity factor is the measure of the energy outputted by an energy system in kWh divided by the total potential energy output. The LCOE is also sensitive to the underlying data, including the capital and operating costs. To minimize these errors, the capacity factor is calculated using the simulation data over the course of 1 year, therefore capturing the near true operation of the system. The equation for LCOE is shown below:

$$\text{LCOE}\left(\frac{\text{€}}{\text{kWh}}\right) = \frac{\sum_{i=1}^{n} C_i + M_i + F_i}{\sum_{i=1}^{n} E_i}$$

where C_i, M_i, and F_i are the total capital, operation/maintenance, and fuel expenditures, respectively. E_i is the total energy outputted from the system. C_i, M_i, and F_i are found through research reasonable values from industrial reports, real installations, and reviewed sources, whereas E_i can be outputted from the simulation model.

Additionally, the internal rate of return (IRR) can be employed to determine the suitability of the VPP investment given the calculated yearly cash flow quantities and can be found by setting the net present value (NPV) equal to zero and solving for the IRR in the equation below:

$$\text{NPV}(0) = \sum_{i=1}^{n}\left(\frac{C_i}{(1+\text{IRR})^i}\right) - C_0$$

Renewable Energy Community VPP Concept Design and Modelling 93

TABLE 4.2

Costing Values for the VPP Components and Lifetime

System	CAPEX (€/kW)	OPEX (€/year)	Lifetime (years)
PV solar	714	1.2% CAPEX	20
Lithium-ion battery	390 (/kWh)	0.3% CAPEX	5 (or 1500 cycles)
PEM fuel cell	2500	2% CAPEX	20
PEM electrolyser	1200		
Hydrogen storage	20 (/kWh)		

where C_i is the cash flow and C_0 is the initial capital for number of years i. The IRR can be used to gauge whether a project spending will overcome the investors hurdle [30], so if the hurdle is 10%, then a value of 10% or higher is required.

The CAPEX and OPEX values used for the calculations are shown in Table 4.2 and collected from various academic and industrial sources, including [19, 31–34].

The environmental impact of the system can be calculated simply by using researched values of the emissions intensity (gCO$_2$e/kWh) of the generation and storage systems in comparison to the grid impact. The grid emissions were found using generation data gathered from Red Electrica de Espana for the year 2020.

$$EI_{total} = \frac{\sum_{j=1}^{m} (EI_j \cdot E_j)}{\sum_{j=1}^{m} (E_j)}$$

where EI_j is the emissions intensity and E_j is the energy output for m number of generators and ESSs. This calculation is performed for each timestep of the simulation to find the dynamic emissions value depending on the instantaneous energy mix of the VPP. The emissions intensity found within literature can vary immensely due to the range of manufacturing techniques and factors considered when performing the life cycle assessment (LCA). For this reason, some values such as the used for the hydrogen system are taken as an educated estimation of the emissions impact based on a variety of sources. The emissions intensity used for each component in the VPP is shown in Table 4.3, with data gathered from [34–37].

TABLE 4.3

Emissions Intensity of the VPP Components

System	Emissions (gCO$_2$e/kWh)
PV solar	45
Lithium-ion battery	73
PEM fuel cell	60
PEM electrolyser and storage tank	60

4.3 RESULTS

This section contains the results from the modelling and simulation of the designed hybrid energy-storage-based multi-building VPP. As mentioned, the key objectives were to present the optimized VPP with integrated hybrid energy system to support the selected Energy Community on the island of Formentera. The first section presents the initial economic optimization of the combined battery and hydrogen system simulated over the course of one year with the available building loads and meteorological data for the year 2020. Based on this assessment, the simulation results of the optimal result are displayed, showcasing the dynamic response of the ESS to the current system load conditions, and excess PV solar energy available. Finally, the environmental impact is studied to determine the emissions improvements made to the system over the reference grid-only use case.

4.3.1 Economic Optimization

Due to model complexity and computational requirements, the optimization process was reduced to a linear process, whereby the battery size was adjusted by a step size of 5 kWh and the hydrogen system remained static. The electrolyser and hydrogen fuel cell are balanced manually to ensure that the hydrogen storage State of Charge starts and finishes the year to the same value of approximately 30%.

The results in Figure 4.12 display the economic assessment for the HESS. The change in lithium-ion battery capacity produces a variation in the LCOE and therefore the IRR. A battery that is undersized for the system created an exponentially increasing LCOE and an IRR with a negative value for capacities smaller than 20 kWh. This occurs because the small battery incurs more cycles per year than a larger battery and, therefore, reduces its lifetime significantly, which worsens for very small sizes. Lithium-ion batteries have been noted to have a lifespan of 5 years

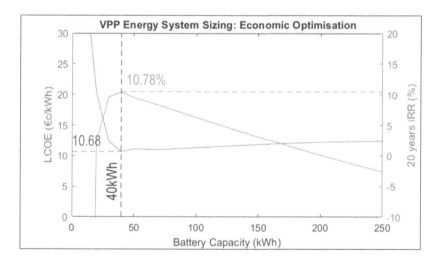

FIGURE 4.12 Economic optimization of the hybrid energy system.

or 1500 cycles (whichever comes first in this study), so a battery that can last the full term of 5 years before requiring a replacement performs much better economically than a small battery. The lowest LCOE occurs at a battery capacity of 50 kWh, before steadily increasing again for large battery sizes. This is because once the balance of cycles and lifespan in years is reached, the larger battery size will naturally come at a higher cost as it would be underutilized in terms of cycles and maintenance costs compound over time. This finding falls in line with other similar studies into grid energy storage options, which also suggest that for large capacities, battery only system begin to be less economically viable compared with the hydrogen system alternatives [38]. It can also be noted for figure that the IRR aligns with the optimal LCOE as can be expected from the calculation, leading to a maximum of 10.78% return of a 20-year system life. It was decided that although a 40-kWh battery provides the optimal solution, a 50-kWh battery with a slightly larger capacity should be used to overcome some of the losses not considered within this report, with only a small difference in LCOE.

Once the system sizing optimization was a complete, an in-depth study of the model simulation can be conducted to determine the system dynamics and response to various loading scenarios through the 1-year period. Table 4.4 displays the output data from the model, including the calculated CAPEX and OPEX for each of the components as well as the combined ESS and total VPP. The results show that overall the CAPEX of the hydrogen system contributes the most to the overall capital. However, because the battery system is limited in lifetime to 5 years, new battery replacements are taken into account within the LCOE calculation. PV solar contributes a relatively high amount to the overall cost but is low when considering the amount of energy generated and expected lifetime, producing an LCOE of 4.7 €cents/kWh, which is in line with expectation from literature [39].

The timeseries results in Figure 4.13 show the dynamic operation of the VPP over the course of one week within the 1-year simulation. On the day of April 3rd for example, the day starts with the fuel cell operating to charge the battery, which is then delivering power to the load. The choppy nature of the battery in charging and discharging is due to the imposed limit of 50% depth of discharge whilst the fuel cell is operating. This is performed to limit the ageing of the battery as it is not able

---

TABLE 4.4
VPP Costing Assessment for Each Component

Component	CAPEX (€)	OPEX (€/year)	Lifetime	Energy/Year (MWh/year)	LCOE (€cents/kWh)	Efficiency (%)
PV solar	80,274.6	1339	20	164.7	4.7	16.5
Battery	19,500	639	5	13.1	21.5	92.5
Hydrogen system	91,000	925	20	46.3	15.1	33.7
ESS total	110,500	1564	20	59.4	17.6	–
VPP total	190,774.6	2903	20	224.1	10.7	24.5

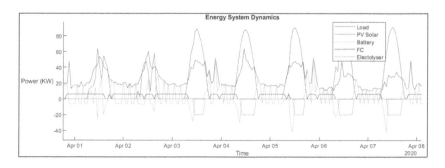

FIGURE 4.13 VPP dynamic response to total system load and PV generation.

to perform a full charge-discharge cycle under these conditions. Once PV solar is available in excess, the battery is fully charged, after which the electrolyser is activated to absorb the excess power for the remainder of the day. As the PV solar power reduces, the battery is reactivated to supply the load, then once fully discharged can be recharged with the fuel cell using the previously generated hydrogen in storage.

It was decided that the fuel cell would be used to charge the battery in a two-stage process rather than a connection directly to the load. This was done for a few of reasons. First, as the fuel cell has the lowest operational efficiency within the system, steps must be taken to ensure that it is operating in its most efficient mode. The efficiency varies greatly over the operating range, with a steep drop at low current densities. Therefore, the fuel cell was set to its nominal power output of 6 kW rather than having the ability of load following and risking a reduction in efficiency. The battery is then charged at that constant rate and delivers to the load with an efficiency of 90%. The second is that a fuel cell that experiences large variations in its operational output could have a significant reduction in lifespan due to degradation of the catalyst membrane [40]. Third, connecting the fuel cell directly to the load would require additional power system components such as an individual inverter and wire connections, increasing overall installation costs. Keeping the ESS as a self-contained system within a single connection to the load would theoretically minimize the complexity of installation.

As noted in the results, this system does not have the capacity to fully satisfy the load at all times of the day. This is due in part to the limited size of PV solar that could be installed on the participating buildings in comparison to their required power demands. This could potentially be remedied by increasing the PV solar capacity on the ground using small-scale solar farms, as building roof space is at its physical limits for this study.

The bar graph is Figure 4.14 shows the relative efficiency of each of the VPP system components. The battery storage system is assumed to have the highest efficiency at 90%, followed by the electrolyser and fuel cell at 83% and 55%, respectively, producing an overall ESS efficiency of ~34%. The hydrogen system naturally has a much lower efficiency than the battery alone model, but this increase is negated when the added cost of the battery and low operation and maintenance costs for the electrolyser and fuel cell systems are considered. The pie chart in figure illustrates the portion of energy delivered by each VPP component to the end-user loads. As mentioned, due to the size of the building loads in comparison to the excess PV solar

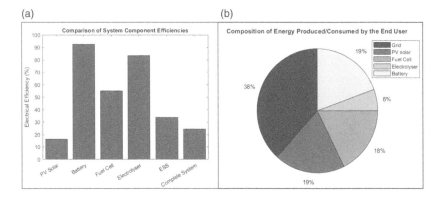

FIGURE 4.14 (a) Efficiency values for each system component and (b) composition of energy produced and consumed.

generation available, a small reliance on the grid connection is still necessary for situations where the hybrid ESS cannot match the given demand, for example during night periods and autumn/winter seasons.

4.3.2 END-USER ENERGY COSTS

The economics of the VPP can be further examined with the analysis of the individual energy costs for each building over the simulation year. It is assumed that all buildings form a multilateral energy agreement in which all end users collectively own the community PV solar installations and the HESS. The energy is then 'traded' between the buildings and the HESS with the aim of reducing the total energy cost for all end users rather than prioritizing a single building. This has the natural effect of benefiting buildings with higher PV solar generation then lower, as more excess energy is available to be used in the system. It can be seen in Figure 4.15 that the energy cost both varies seasonally throughout the year, and between buildings included within the VPP.

The reason for lower energy costs during the summer period is simply because the increase in sun peak hours (SPH) during this season has allowed for higher energy generation from the solar panels. As expected from the energy sharing system, the buildings with the larger PV solar system have a lower average cost throughout the year. For example, building 3 (school) has the second-largest solar array after building 4 (football field) and a slightly lower average load compared to the latter. This has led to the best LCOE, whereas building 1 with the smallest PV solar array and a relatively high load has the highest LCOE, because the PV solar has the lowest single LCOE of all components within the VPP.

4.3.3 ENVIRONMENTAL IMPACT ASSESSMENT

The average emissions factor calculated for the grid activity on the island is between approximately 350 gCO_2e/kWh for the year of 2020, which is relatively low compared

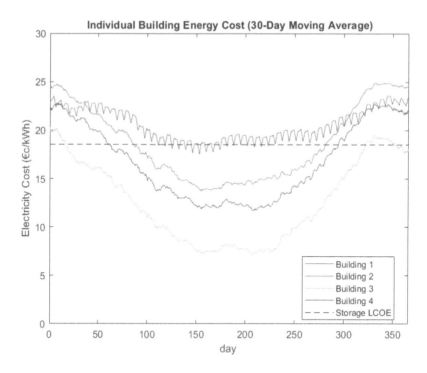

FIGURE 4.15 Individual Energy Community building cost throughout simulation year (30-day moving average).

with other years. This is due in part to the COVID-19 pandemic reducing the energy demand in Spain, which in term allowed a higher relative usage of renewables rather than relying on fossil fuel power stations to supply power to the grid. Closed Cycle Gas Turbine (CCGT) and coal power plants are cited to produce 490 and 820 gCO2e/kWh [35], respectively, which is much higher than the calculated grid impact due to the renewable energy generation within the energy mix.

Figure 4.16 shows the 30-day moving average of the grid emissions intensity in comparison to the four buildings within the VPP. With an emissions intensity of 45 g, the presence of PV solar within the energy mix has the largest overall impact in the reduction of environmental emissions. With the ability to store the excess PV solar energy, the low carbon generation can be utilized more effectively by increasing self-consumption, further reducing emissions. Like the LCOE values, the emissions produced are lowest during the summer months due to the increase in solar hours.

4.3.4 VPP Analysis

Overall, the designed VPP system was successfully able to reduce both the energy cost and emissions impact of the select Energy Community of buildings when compared to the grid alternative. This was achieved by maximizing the self-consumption of PV solar generation with the optimized hybrid ESS.

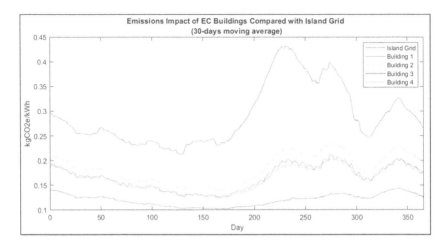

FIGURE 4.16 Emissions impact of EC buildings compared with the island grid reference (30-day moving average).

The data in Table 4.5 shows the percentage improvements in emissions and cost for the four buildings. The improvements in average emissions range from 42.8% for building 1 to 58.3% for building 3 when compared to the grid reference. As with the previously discussed results, this data illustrates the relationship between increased PV solar generation and decreased emissions production from the VPP for each building. Similarly, building 3 produced the best improvement in energy cost in comparison to the grid reference at 49.7% or 15.1 €cent/kWh. The low LCOE has also derived from the optimized sizing of the hybrid energy system, with a combined cost of 17.6 €cent/kWh.

4.4 CONCLUSION

The principal objective laid out in this work was to design a suitable small-scale community-based VPP system utilizing real production and consumption data from the field. The produced VPP integrated four community-operated buildings on the

TABLE 4.5
Benefit Analysis of the Community Buildings Within the VPP, Within Improvements in Cost and Environmental Impact

Site	Average Emissions (kgCO$_2$e/kWh)	Improvement (%)	Average Cost (€cent/kWh)	Improvement (%)
Building 1	0.166	42.8	22.6	24.7
Building 2	0.184	36.6	21.3	29.0
Building 3	0.121	58.3	15.1	49.7
Building 4	0.164	43.4	19.6	34.7

small Mediterranean island of Formentera for the purpose of maximizing the self-consumption of the installed PV solar arrays. The solar capacity was artificially increased based on spatial and geometric studies performed on the roofing area using the commercially available Helioscope software. This was done to highlight more clearly the benefits of increased renewables within the VPP.

The next objective was to implement a hybrid ESS combining the outputs of a lithium battery and regenerative hydrogen system to balancing the output of the combined PV solar system and community building loads. This objective was completed by producing a dynamic model that could simulate the time-based response of the system to the various consumption and generation scenarios throughout 1 year. Each component was modelled individually and could be tuned with parameterization to match the performance requirements of the real system. The novel energy management system design implements a state-based logic control to prioritize the most efficient operating mode for the VPP at a given time and based on the quantity of excess PV solar available.

The final objective was to optimize system sizing and maximize the economic benefits, including the LCOE, and demonstrate reductions in the environmental impact. This was completed by varying the hybrid ESS size in a linear optimization routine to find the best combination of lithium-ion battery and hydrogen system sizing. The optimal result gave an LCOE value of 10.7 €cent/kWh for the combined VPP, which compared to an assumed grid cost of ~30 €cent/kWh displays a large improvement and potential for significant cost savings over the system lifetime. In an analysis of the individual buildings performance, the results showed a heavy correlation with the PV solar generation available, due to the lower energy cost and more flexibility delivered to the shared VPP. The school building with the largest solar array reached improvements of 58.3% and 49.7% in yearly cost and environmental emissions, respectively. The system IRR of 10.78% is relatively low to be considered commercially viable for this size of system, so in future works more research should be conducted into other auxiliary services that the VPP could provide to the grid, such as frequency balancing and integration with the energy spot market. Furthermore, a concrete business model that considers these and other VPP services should be presented to make the system a more attractive prospect for end users.

ACKNOWLEDGEMENT

This work has been carried out in the framework of the European Union's Horizon 2020 research and innovation program under grant agreement No 957852 (Virtual Power Plant for Interoperable and Smart isLANDS – VPP4ISLANDS).

REFERENCES

1. Renewables 2020 Global Status Report, REN21 Secretariat, Paris, 2020.
2. United Nations, The Paris Agreement, United Nations, Paris, France, 2015.
3. United Nations Environment Programme, Emissions Gap Report 2021: The Heat Is On – A World of Climate Promises Not Yet Delivered, UNEP, Nairobi, 2021.

4. IPCC, Climate Change 2021: The Physical Science Basis. Contribution of Working Group I to the Sixth Assessment Report of the Intergovernmental Panel on Climate Change, Cambridge University Press, Cambridge, UK, 2021.
5. UK National Statistics, UK Energy in Brief, Department for Business, Energy & Industrial Strategy, London, UK, 2021.
6. BP, Statistical Review of World Energy 69th Edition, British Petroleum, London, UK, 2020.
7. IRENA, Global Energy Transformation: A Roadmap to 2050, International Renewable Energy Agency, Abu Dhabi, 2018.
8. Deloitte, Supercharged: Challenges and Opportunities in Global Battery Storage Markets, Deloitte Development LLC, Washington, DC, 2018.
9. S. Ghavidel, L. Li, J. Aghaei, Y. Tao and J. Zhu, "A Review on the Virtual Power Plant: Components and Operation Systems," in 2016 IEEE International Conference on Power System Technology (POWERCON), Wollongong, NSW Australia, 2016.
10. M. Braun, "Virtual Power Plants in Real Applications-Pilot Demonstrations in Spain and England as Part of the European Project FENIX," in ETG-Fachbericht-Internationaler, ETG-Kongress, Düsseldorf, 2009.
11. C. Binding, D. Gantenbein, B. Jansen, O. Sundstrom, P. Andersen, F. Marra, B. Poulsen and C. Træholt, "Electric Vehicle Fleet Integration in the Danish EDISON Project – A Virtual Power Plant on the Island of Bornholm," in IEEE PES General Meeting, Minneapolis, MN, 2010.
12. Con Edison, REV Demonstration Project Outline, Con Edison, New York, NY, 2015.
13. Sonnen Community, "Sonnen," 2021. [Online]. Available: https://sonnengroup.com/sonnencommunity/. [Accessed 25 June 2021].
14. AGL, Virtual Power Plant in South Australia, AGL, Adelaide, Australia, 2017.
15. A. Ferrario, A. Bartolini, F. Manzano, F. Vivas, G. Comodi, S. McPhail and J. Andujar, "A Model-Based Parametric and Optimal Sizing of a Battery/Hydrogen Storage of a Real Hybrid Microgrid Supplying a Residential Load: Towards Island Operation," *Advances in Applied Energy*, vol. 3, p. 100048, 2021.
16. K. Knosala, L. Kotzur, F. Roben, P. Stenzel, L. Blum, M. Robinius and D. Stolten, "Hybrid Hydrogen Home Storage for Decentralized Energy Autonomy," *International Journal of Hydrogen Energy*, Vols. 21748-21763, p. 46, 2021.
17. D. Emad, M. El-Hameed and A. El-Fergany, "Optimal Techno-Economic Design of Hybrid PV/Wind System Comprising Battery Energy Storage: Case Study for a Remote Area," *Energy Conversion and Management*, vol. 249, p. 114847, 2021.
18. M. Abdolrasol, A. Mohamed and M. Hannan, "Virtual Power Plant and Microgrids Controller for Energy Management Based on Optimization Techniques," *Journal of Electrical Systems*, vol. 13, no. 2, pp. 285–294, 2017.
19. G. Jansen, Z. Dehouche and H. Corrigan, "Cost-Effective Sizing of a Hybrid Regenerative Hydrogen Fuel Cell Energy Storage System for Remote & Off-Grid Telecom Towers," *International Journal of Hydrogen Energy*, vol. 46, no. 22, 2021.
20. European Commission, *Prices and Costs of EU Energy*, Ecofys, Netherlands, 2016.
21. Eurostat, "Greenhouse Gas Emissions per Capita," *Eurostat*, 2020.
22. A. Fathy, D. Yousri, T. Alanazi and H. Rezk, "Minimum Hydrogen Consumption Based Control Strategy of Fuel Cell/PV/Battery/Supercapacitor Hybrid System Using Recent Approach Based Parasitism-Predation Algorithm," *Energy*, vol. 225, p. 120316, 2021.
23. I. M. Syed and A. Yazdani, "Simple Mathematical Model of Photovoltaic Module for Simulation in Matlab/Simulink," in 2014 IEEE 27th Canadian Conference on Electrical and Computer Engineering (CCECE), Toronto, Ontario, Canada, 2014.
24. M. King, D. Li, M. Dooner, S. Ghosh, J. N. Roy, C. Chakraborty and J. Wang, "Mathematical Modelling of a System for Solar PV Efficiency Improvement Using Compressed Air for Panel Cleaning and Cooling," *Energies*, vol. 14, no. 4072, pp. 1–18, 2021.

25. S. Motahhir, e. g. Abdelaziz and A. Derouich, "Shading Effect to Energy Withdrawn from the Photovoltaic Panel and Implementation of DMPPT Using C Language," *International Review of Automatic Control*, vol. 9, no. 2, p. 88, 2016.
26. Q. Wei, Functional Nanofibers and their Applications, Woodhead Publishing, Cambridge, UK, 2012.
27. O. Tremblay and L.-A. Dessaint, "Experimental Validation of a Battery Dynamic Model for EV Applications," World Electric Vehicle Journal, vol. 3, pp. 289–298, 2009.
28. D. Matthias, C. Andrew, A. Sinclair and J. McDonald, "Dynamic Model of a Lead Acid Battery for Use in a Domestic Fuel Cell System," Journal of Power Sources, vol. 161, no. 2, pp. 1400–1411, 2006.
29. C. Shepard, "Design of Primary and Secondary Cells – Part 2: An Equation Describing Battery Discharge," Journal of Electrochemical Society, vol. 112, pp. 657–664, 1965.
30. A. Eltamaly and M. Mohamed, Advances in Renewable Energies and Power Technologies: Volume 2: Biomass, Fuel Cells, Geothermal Energies, and Smart Grids, Elsevier, 2018, pp. 231–313.
31. W. Cole and W. Frazier, Cost Projections for Utility-Scale Battery Storage, National Renewable Energy Laboratory, Golden, CO, 2019.
32. T. Aquino, M. Roling, C. Baker and L. Rowland, Battery Energy Storage Technology Assessment, Platte River Power Authority, Fort Collins, Colorado, 2017.
33. S. Kharel and B. Shabani, "Hydrogen as a Long-Term Large Scale Energy Storage Solution to Support Renewables," *Energies*, vol. 11, p. 2825, 2018.
34. IRENA, Hydrogen From Renewable Power: Technology Outlook for the Energy Transition, International Renewable Energy Agency, Abu Dhab, 2018.
35. O. Edenhofer, R. Pichs-Madruga, Y. Sokona, K. Seyboth, P. Matschoss, S. Kadner, T. Zwickel, P. Eickemeier, G. Hansen, S. Schloemer and C. von Stechow, IPCC Special Report on Renewable Energy Sources and Climate Change Mitigation, Cambridge University Press, Cambridge, UK, 2011.
36. H. Melin, Analysis of the Climate Impact of Lithium-Ion Batteries and How to Measure It," Circular Energy Storage, London, UK, 2019.
37. L. Usai, C. Hung, F. Vasquez, M. Windsheimer, O. Burheim and A. Strømman, "Life Cycle Assessment of Fuel Cell Systems for Light Duty Vehicles, Current State-of-the-Art and Future Impacts," *Journal of Cleaner Production*, vol. 280, p. 125086, 2021.
38. M. Pellow, C. Emmott, C. Barnhart and S. Benson, "Hydrogen or Batteries for Grid Storage? A Net Energy Analysis," *Energy Environ. Sci (RSoC)*, vol. 8, p. 1938, 2015.
39. N. A. Ludin, A. A. N. Affandi, K. Purvis-Roberts, A. Ahmad, A. M. Ibrahim, K. Sopian and S. Jusoh, "Environmental Impact and Levelised Cost of Energy Analysis of Solar Photovoltaic Systems in Selected Asia Pacific Region: A Cradle-to-Grave Approach," *Sustainability*, vol. 13, no. 1, p. 396, 2021.
40. M. Mayur, M. Gerard, P. Schott and W. Bessler, "Lifetime Prediction of a Polymer Electrolyte Membrane Fuel Cell under Automotive Load Cycling Using a Physically-Based Catalyst Degradation Model," *Energies*, vol. 11, no. 8, p. 2054, 2018.

5 A Comprehensive Smart Energy Management Strategy for TVPP, CVPP, and Energy Communities

Ehsan Heydarian-Forushani[1,2] and
Seifeddine Ben Elghali[1]
[1]Laboratory of Information & Systems (LIS-UMR CNRS 7020), Aix-Marseille University, Marseille, France
[2]Department of Electrical and Computer Engineering, Qom University of Technology, Qom, Iran.

CONTENTS

5.1 Introduction 105
5.2 Objective 105
5.3 Mathematical Model of Optimization Engine in
     Different Operation Modes.. 105
    5.3.1 TVPP Model 105
    5.3.2 CVPP Model.. 109
    5.3.3 Energy Community Model 109
5.4 Mathematical Model of Components Within the VPP............................. 111
    5.4.1 DG Units.. 111
    5.4.2 ES Units... 111
    5.4.3 Flexible Loads 112
5.5 Numerical Analysis and Discussions.. 112
Acknowledgement 116
References... 117

NOMENCLATURES

INDICES

b, b'	system buses
i	conventional diesel units
es	energy storage units
pv	PV generation sites
Sb	source buses
fl	flexible loads

DOI: 10.1201/9781003257202-6

104 Virtual Power Plant Solution for Future Smart Energy Communities

wf	wind farms
t	time slots
w	scenarios
m	linear partitions in load flow linearization

PARAMETERS

$\lambda_{w,t}^{DA}$	day-ahead market price scenarios (\$/kWh)
λ_t^{Grid}	upstream market price (\$/kWh)
C_i^{DG}	generation cost of DG units (\$/kWh)
SUC_i^{DG}	start-up cost of DG units (\$)
INC_{fl}	incentive payment for load curtailment (\$/kWh)
$P_{pv,w,t}^{PV}$	PV site generation scenarios (kW)
$P_{wf,w,t}^{Wind}$	wind farm generation scenarios (kW)
$P_{b,w,t}^{D}$	active power consumption scenarios (kW)
$R_{b,b'}/X_{b,b'}/Z_{b,b'}$	impedance elements of branches (Ohm)
$Q_{b,w,t}^{D}$	reactive power consumption scenarios (kVar)
$P_i^{DG,\max}$	maximum DG capacity limit for active power (kW)
$P_i^{DG,\min}$	minimum DG capacity limit for active power (kW)
$P_{es}^{DchES,\max}/P_{es}^{ChES,\max}$	maximum discharging/charging power of ES (kW)
α_{fl}^{flex}	percentage of flexible load at each load point (%)
$V_b^{\min}/V_b^{\max}$	minimum/maximum allowable voltage magnitude of buses (kV)
RU_i^{DG}/RD_i^{DG}	ramp-up/down capability of DG units (kW/h)
MUT_i/MDT_i	minimum up/down time of DG units (h)
LR_{fl}^{pickup}	pickup rate of flexible loads (kW/h)
LR_{fl}^{drop}	drop-off rate of flexible loads (kW/h)
$\eta_{Ch}^{ES}/\eta_{Dch}^{ES}$	charging/discharging efficiency of ESs
$SOE_{es}^{ES,\min}/SOE_{es}^{ES,\max}$	minimum/maximum energy limit of ESs (kWh)
ρ_w	occurrence probability of scenarios

VARIABLES

$P_{Sb,t}^{DA}$	the amount of selling or purchasing power in DA energy market (kW)
P_t^{Grid}	the amount of selling or purchasing power from upstream grid (kW)
$P_{i,w,t}^{DG}$	scheduled active output power of DG units (kWh)
$P_{fl,w,t}^{flex}$	curtailed active amount of flexible load (kWh)
$P_{es,w,t}^{ChES}/P_{es,w,t}^{DchES}$	scheduled charge/discharge power of ESs (kW)
$\left|V_{b,w,t}\right|$	voltage magnitude of network buses (kV)
$I_{b,b',w,t}$	current of network branches (A)
$Q_{i,w,t}^{DG}$	scheduled reactive output power of DG units (kVarh)
$Q_{fl,w,t}^{flex}$	curtailed reactive amount of flexible load (kVarh)
$P_{b,b',w,t}^{+}/P_{b,b',w,t}^{-}$	auxiliary variables for active power flow (kW)
$Q_{b,b',w,t}^{+}/Q_{b,b',w,t}^{-}$	auxiliary variables for reactive power flow (kVar)

$U_{i,t}^{DG}$	binary variable indicates on/off situation of DG units
$U_{es,t}^{ES\_Ch}$	binary variable indicates charge/discharge situation of ES units
$SOE_{es,w,t}^{ES}$	state of the energy level of ESs (kWh)

5.1 INTRODUCTION

In the context of future smart grids, the distributed and small-scale energy resources will play a more remarkable role in comparison with their present situation. One of the main technical and commercial challenges for integration of distributed energy resources (DERs) into the grid is their invisibility from system operator's viewpoint. Virtual power plant (VPP) concept seems an appropriate solution to tackle such a problem. In fact, a VPP is composed of small DERs (traditional and renewables), controllable loads, and energy storage (ES) devices and can coordinate them in order to provide several local and system services through participation in different electricity markets with the aim of maximizing its profits.

5.2 OBJECTIVE

As mentioned before, the VPP aggregates and manages not only distributed generations such as diesel generators (DGs), photovoltaic (PVs), and wind turbines (WTs) but also ESs as well as flexible loads (FLs) through an integrated optimization framework so that there is a maximization of its revenues in various electricity markets. In other words, a VPP is a flexible representation of a portfolio of DERs that can be used to make contracts in the wholesale market and offer services to the system operator. On this basis, the objective of this deliverable is to develop a centralized optimization engine for a VPP that can be operated in different modes, including technical VPP (TVPP) mode, commercial VPP (CVPP) mode, and also energy community (EC) mode according to the user preferences.

5.3 MATHEMATICAL MODEL OF OPTIMIZATION ENGINE IN DIFFERENT OPERATION MODES

5.3.1 TVPP MODEL

A TVPP is a type of VPP that consists of DERs from the same geographic location. The TVPP encompasses the real-time impacts of the local network on the aggregated profile of DERs as well as representing the cost and operating constraints of the portfolio. The TVPP can participate in day-ahead and intraday energy markets for purchasing or selling power on behalf of the existing resources under its umbrella. The operator of a TVPP, so-called VPP aggregator, requires detailed information of the local network as well as the existing resources [1]. In fact, in this case, the optimization engine must be considered the network structure and run an optimal power flow in order to avoid either feeder's congestion or voltage magnitude violation.

To model the stochastic nature of renewable resources as well as taking into account the electricity market price's volatility, the proposed methodology makes use of a two-stage stochastic programming framework and considers a network-constrained market participation procedure from a TVPP point of view. The structuring of the stochastic programming problem into two stages is justified by the fact that the TVPP aggregator must submit its offers into different electricity markets, while facing a number of uncertainties such as volatility in market prices, load consumption, and generation of renewable resources. The proposed model has a hierarchical structure so that it encompasses both day-ahead and intraday time frames as can be observed in Figure 5.1.

In fact, the optimal strategy of VPP for participation in day-ahead energy market is determined at the first stage, while the second stage is assigned for intraday market decisions which are based on the updated information obtained between the closure of the day-ahead and the intraday energy markets.

The objective function of the TVPP is formulated in Eq. (5.1). The objective function is the total expected cost during the scheduling horizon, which contains three terms, including cost/revenue in day-ahead energy market, cost of DG power production, and incentive payment to FL, respectively. It is notable that the objective function that must be optimized for the second stage is completely similar to Eq. (5.1) with the difference that the power traded in day-ahead market is an input parameter since the day-ahead market has been cleared. In fact, the first term in the objective function must be substituted by the revenue or cost term for selling or purchasing power in intraday market.

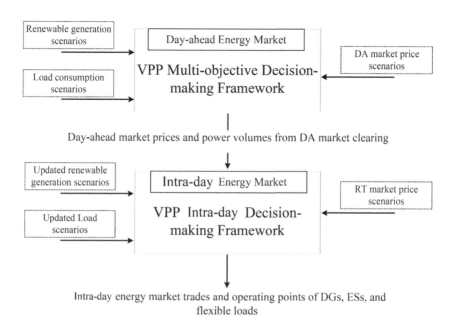

FIGURE 5.1 Schematic of the proposed hierarchical optimization model.

A Comprehensive Smart Energy Management Strategy

$$Min \quad OF = \sum_{w=1}^{NW} \rho_w \left(F_w^1 + F_w^2 + F_w^3 \right)$$

$$F_w^1 = \sum_{Sb=1}^{NSb} \sum_{t=1}^{NT} \lambda_{w,t}^{DA} P_{Sb,t}^{DA}$$

$$F_w^2 = \sum_{t=1}^{NT} \sum_{i=1}^{NDG} \left[C_i^{DG} \left(P_{i,t}^{DG} + \Delta P_{i,w,t}^{DG} \right) + C\_Startup_{i,t} \right] \tag{5.1}$$

$$F_w^3 = \sum_{t=1}^{NT} \sum_{fl=1}^{NFL} INC_{fl} P_{fl,w,t}^{flex}$$

The active and reactive power balance constraints have been formulated in Eqs. (5.2) and (5.3), respectively. Note that the first term in Eq. (5.2) is associated with power transactions with the upstream grid and just must be taken into account for the source bus.

$$\sum_{Sb \in b} P_{Sb,t}^{DA} + \sum_{i \in b} P_{i,w,t}^{DG} + \sum_{pv \in b} P_{pv,w,t}^{PV} + \sum_{wf \in b} P_{wf,w,t}^{Wind} + \sum_{es \in b} \left(P_{es,w,t}^{Dch} - P_{es,w,t}^{Ch} \right)$$

$$+ \sum_{fl \in b} P_{fl,w,t}^{flex} + \sum_{b'} \left(P_{b',b,w,t}^{+} - P_{b',b,w,t}^{-} \right) \tag{5.2}$$

$$- \sum_{b'} \left[\left(P_{b,b',w,t}^{+} - P_{b,b',w,t}^{-} \right) + R_{b,b'} I2_{b,b',w,t} \right] = P_{b,w,t}^{D}$$

$$\sum_{Sb \in b} Q_{Sb,t}^{DA} + \sum_{i \in b} Q_{i,w,t}^{DG} + \sum_{fl \in b} Q_{fl,w,t}^{flex}$$

$$+ \sum_{b'} \left(Q_{b',b,w,t}^{+} - Q_{b',b,w,t}^{-} \right) \tag{5.3}$$

$$- \sum_{b'} \left[\left(Q_{b,b',w,t}^{+} - Q_{b,b',w,t}^{-} \right) + X_{b,b'} I2_{b,b',w,t} \right] = Q_{b,w,t}^{D}$$

The active and reactive power relations are considered according to the power factor concept as can be observed in the following equation.

$$Q_{b,w,t}^{D} = PF \times P_{b,w,t}^{D}$$

$$Q_{Sb,t}^{DA} = PF \times P_{Sb,t}^{DA} \tag{5.4}$$

$$Q_{fl,w,t}^{flex} = PF \times P_{fl,w,t}^{flex}$$

The TVPP must consider the power flow restrictions within its optimization model. On this basis, the linear form of load flow equations has been formulated in Eqs. (5.5)–(5.15) [2].

$$V2_{b,w,t} - 2R_{b,b'}\left(P_{b,b',w,t}^{+} - P_{b,b',w,t}^{-}\right) + X_{b,b'}\left(Q_{b,b',w,t}^{+} - Q_{b,b',w,t}^{-}\right) - Z2_{b,b'}\,I2_{b,b',w,t}$$
$$- V2_{b',w,t} = 0 \tag{5.5}$$

$$V2_{b,w,t}^{Rated}\,I2_{b,b',w,t} = \sum_{m=1}^{NM}\left[(2m-1)\Delta S_{b,b',w,t}\,\Delta P_{b,b',m,w,t}\right]$$
$$+ \sum_{m=1}^{NM}\left[(2m-1)\Delta S_{b,b',w,t}\,\Delta Q_{b,b',m,w,t}\right] \tag{5.6}$$

$$P_{b,b',w,t}^{+} + P_{b,b',w,t}^{-} = \sum_{m=1}^{NM}\Delta P_{b,b',m,w,t} \tag{5.7}$$

$$Q_{b,b',w,t}^{+} + Q_{b,b',w,t}^{-} = \sum_{m=1}^{NM}\Delta Q_{b,b',m,w,t} \tag{5.8}$$

$$0 \le \Delta P_{b,b',m,w,t} \le \Delta S_{b,b',w,t} \tag{5.9}$$

$$0 \le \Delta Q_{b,b',m,w,t} \le \Delta S_{b,b',w,t} \tag{5.10}$$

$$\Delta S_{b,b',w,t} = \left(V^{Rated}\,I_{b,b'}^{max}\right)/NM \tag{5.11}$$

$$0 \le I2_{b,b',w,t} \le \left|I_{b,b'}^{max}\right|^{2} \tag{5.12}$$

$$0 \le P_{b,b',w,t}^{+} + P_{b,b',w,t}^{-} \le V^{Rated}\,I_{b,b'}^{max} \tag{5.13}$$

$$0 \le Q_{b,b',w,t}^{+} + Q_{b,b',w,t}^{-} \le V^{Rated}\,I_{b,b'}^{max} \tag{5.14}$$

$$\left|V_{b}^{min}\right|^{2} \le V2_{b,w,t} \le \left|V_{b}^{max}\right|^{2} \tag{5.15}$$

Equation (5.5) is considered with the aim of balancing voltage between two nodes. It is notable that $V2$ in Eq. (5.5) is an auxiliary variable that shows the linear form of squared voltage relation. Also, linearization of active and reactive power flows that appear in the apparent power is formulated in Eq. (5.6). Equations (5.7)–(5.11) have been formulated for piecewise linearization. The number of blocks needed to linearize the quadratic curve is considered to be five based on [3], which maintains

A Comprehensive Smart Energy Management Strategy

the right balance between accuracy and computational requirements. Further descriptions and justifications of the network model used can be found in [4, 5]. The maximum allowable current flow of branches is taken into account in Eq. (5.12). Note that $I2$ refers to an auxiliary variable that demonstrates linear form of the squared current flow I^2 in a given branch. Moreover, at most one of these two positive auxiliary variables, i.e., $P_{b,b',w,t}$ and $Q_{b,b',w,t}$, can be nonzero in a time. This condition is again implicitly enforced by optimality. Equations (5.13) and (5.14) restrict these variables through the maximum apparent power for completeness. Finally, Eq. (5.15) represents the allowable voltage magnitude at each node. The remaining constraints are associated with the existing resources within the TVPP, which are presented in the following.

5.3.2 CVPP MODEL

CVPP performs commercial aggregation and does not take into account any network operation aspects that active distribution networks have to consider for stable operation [6]. The aggregated DER units are not necessarily constrained by location but can be distributed throughout different distribution grids. Hence, a single distribution network region may have more than one CVPP-aggregating DER units in its region. The objective function of CVPP is completely similar to Eq. (5.1); however, the active power balance constraint is different due to non-consideration of network. Also, there is no need to consider reactive power in the case of CVPP due to the same reason. The active power balance constraint for CVPP is modelled in Eq. (5.16).

$$P_t^{DA} + \sum_i P_{i,w,t}^{DG} + \sum_{pv} P_{pv,w,t}^{PV} + \sum_{wf} P_{wf,w,t}^{Wind} + \sum_{es} \left(P_{es,w,t}^{Dch} - P_{es,w,t}^{Ch} \right) + \sum_{fl} P_{fl,w,t}^{flex} = P_{w,t}^{D} \quad (5.16)$$

The remaining constraints are assigned to the existing resources within the CVPP, which are presented in the following.

5.3.3 ENERGY COMMUNITY MODEL

EC is an innovative concept that has been developed to facilitate grid integration of small-scale renewable energy resources, optimize the consumption pattern of customers, and alleviate the loading of the grid through using available flexibility of active prosumers. In the following, a general model of an EC consisting of several small-scale resources is given. It is notable that these resources are spatially very close to each other and connected to the same distribution network. Also, it is assumed that there is a non-profit community coordinator who dispatches the existing resources within the community and manages the transactions with upstream utility as well as other ECs in order to supply the load with minimum cost. The conceptual schematic of the considered model is presented in Figure 5.2.

As it can be observed in Figure 5.2, the EC coordinator runs the local market in order to find an optimal dispatch for the components within the EC, including PV generation, FLs, DGs, and ES units. In this regard, the ECs can also have transactions with other ECs in order to share their available flexibility with each other.

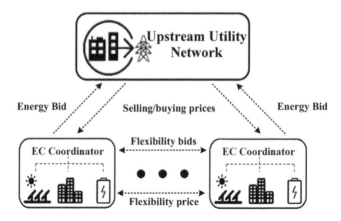

FIGURE 5.2 Conceptual schematic of energy community model.

Moreover, the ECs can buy or sell power to the upstream utility with the aim of ensuring power balance within the community. The mathematical formulation of EC is presented later.

The objective function is a cost minimization problem that has been formulated in Eq. (5.17). The first and second terms in the first line of the objective function represent the costs/revenues as a result of transactions with the upstream utility. The first term in the second line of Eq. (5.17) associates with the cost of DG's power generation containing their start-up cost. Finally, the last term of objective function is assigned to incentive payments for load curtailment through FLs.

$$\text{Min} \quad OF = \Delta \times \sum_{t=1}^{NT} \lambda_t^{Grid} P_t^{Grid}$$
$$+ \sum_{t=1}^{NT} \sum_{w=1}^{NW} \rho_w \left\{ \sum_{i=1}^{NDG} \left[\Delta \times C_i^{DG} P_{i,w,t}^{DG} + C\_Startup_{i,t} \right] \right.$$
$$\left. + \Delta \times \sum_{fl=1}^{NFL} INC_{fl} P_{fl,w,t}^{flex} \right\} \quad (5.17)$$

The Δ in the previous equation represents the time step coefficient and can be defined as $\Delta = Time\_step/60$. The $Time\_step$ in fact indicates the resolution of optimization time slots in minutes. The previous objective function should be minimized while satisfying a number of constraints for different devices. The EC power balance constraint has been modelled in Eq. (5.18).

$$P_t^{Grid} + \sum_{pv=1}^{NPV} P_{pv,w,t}^{PV} + \sum_{i=1}^{NDG} P_{i,w,t}^{DG} + \sum_{fl=1}^{NFL} P_{fl,w,t}^{flex} + \sum_{es=1}^{NES} \left(P_{es,w,t}^{DchES} - P_{es,w,t}^{ChES} \right) = P_{w,t}^{D} \quad (5.18)$$

A Comprehensive Smart Energy Management Strategy

5.4 MATHEMATICAL MODEL OF COMPONENTS WITHIN THE VPP

This section is assigned to the model of existing resources within the VPP and is similar for all operation modes, i.e., TVPP, CVPP, or EC. The considered components are DGs, ESs, and FLs that are modelled in this section. Note that the renewable generations such as PV and WT have been taken into account. According to the priority of renewable resources in dispatch, these resources are modelled to such a negative demand. In fact, the whole production of these resources is integrated into the VPP power scheduling. Moreover, the operation costs of the mentioned resources assume to be zero that is a logical assumption.

5.4.1 DG UNITS

The constraints in Eqs. (5.19)–(5.25) are related to technical restrictions of DGs. The minimum and maximum ranges of output power of DGs are shown in Eq. (5.19). The ramp-up and ramp-down capabilities of conventional DGs are formulated in Eqs. (5.20) and (5.21), separately. Equation (5.22) is assigned to start-up cost of DGs. Moreover, Eqs. (5.23) and (5.24) indicate minimum up and down time limitations of DGs, respectively.

$$P_i^{DG,\min} U_{i,t}^{DG} \leq P_{i,w,t}^{DG} \leq P_i^{DG,\max} U_{i,t}^{DG} \tag{5.19}$$

$$P_{i,w,t+1}^{DG} - P_{i,w,t}^{DG} \leq \Delta \times RU_i^{DG} \tag{5.20}$$

$$P_{i,w,t}^{DG} - P_{i,w,t+1}^{DG} \leq \Delta \times RD_i^{DG} \tag{5.21}$$

$$
\begin{aligned}
& C\_Startup_{i,t} \geq SUC_i \times \left(U_{i,t}^{DG} - U_{i,t-1}^{DG} \right) \quad for\, t \succ 1 \quad and \\
& C\_Startup_{i,t} \geq SUC_i \times U_{i,t}^{DG} \quad for\, t = 1 \\
& C\_Startup_{i,t} \geq 0
\end{aligned}
\tag{5.22}$$

$$\sum_{t'=t}^{t+MUT_i-1} \left(1 - U_{i,t'}^{DG}\right) + MUT_i \ \left(U_{i,t}^{DG} - U_{i,t-1}^{DG} \right) \leq MUT_i \tag{5.23}$$

$$\sum_{t'=t}^{t+MDT_i-1} U_{i,t'}^{DG} + MDT_i \ \left(U_{i,t-1}^{DG} - U_{i,t}^{DG} \right) \leq MDT_i \tag{5.24}$$

5.4.2 ES UNITS

The constraints of ESs that should be satisfied are formulated in Eqs. (5.25)–(5.30). Limitations on charging and discharging power of ESs are modelled in Eqs. (5.25) and (5.26). The state of energy (SOE) level and its limits are also given in Eqs. (5.27)

112   Virtual Power Plant Solution for Future Smart Energy Communities

and (5.28), separately. Furthermore, it is presumed that the SOE level at the end of scheduling period must be greater or equal to the amount of SOE at the beginning of scheduling period as formulated in Eq. (5.29). This is due to the fact that the ES units must have sufficient energy for the next day.

$$0 \leq P_{es,w,t}^{ChES} \leq \Delta \times P_{es}^{ChES,\max} \times U_{es,t}^{ES\_Ch} \tag{5.25}$$

$$0 \leq P_{es,w,t}^{DchES} \leq \Delta \times P_{es}^{DchES,\max} \times \left(1 - U_{es,t}^{ES\_Ch}\right) \tag{5.26}$$

$$SOE_{es,w,t}^{ES} = SOE_{es,w,t-1}^{ES} + \eta_{Ch}^{ES} P_{es,w,t}^{ChES} - P_{es,w,t}^{DchES} / \eta_{Dch}^{ES} \tag{5.27}$$

$$SOE_{es}^{ES,\min} \leq SOE_{es,w,t}^{ES} \leq SOE_{es}^{ES,\max} \tag{5.28}$$

$$SOE_{es,w,t}^{ES}\big|_{t=NT} \geq SOE_{es}^{ES,ini} \tag{5.29}$$

5.4.3 FLEXIBLE LOADS

The other set of constraints is associated with FLs as shown in Eqs. (5.30)–(5.32). Equation (5.30) restricts the amount of FL due to the fact that just a portion of the load is flexible. Also, the maximum load-pickup and load-drop rates for flexible demand are stated in Eqs. (5.31) and (5.32), respectively.

$$0 \leq P_{fl,w,t}^{flex} \leq \alpha_{fl}^{flex} P_{w,t}^{D} \tag{5.30}$$

$$P_{fl,w,t}^{flex} - P_{fl,w,t-1}^{flex} \leq \Delta \times LR_{fl}^{pickup} \tag{5.31}$$

$$P_{fl,w,t-1}^{flex} - P_{fl,w,t}^{flex} \leq \Delta \times LR_{fl}^{drop} \tag{5.32}$$

5.5 NUMERICAL ANALYSIS AND DISCUSSIONS

In order to show the effectiveness of the proposed model, some numerical analyses have been reported in this section. To this end, a CVPP is considered with the following assets. There is a 500-kW wind farm and a 200-kW PV site so that each of the wind farm and PV site generation is modelled considering three independent scenarios, including as forecast, high, and low, with probabilities 0.6, 0.2, and 0.2, respectively. Also, there is one ES unit with the energy capacity of 200 kWh and maximum charging/discharging rates of 100 kW/h. The charge and discharge efficiencies of ES are assumed to be 85%. In addition, it is assumed that the maximum

TABLE 5.1
DGs Technical and Cost Data

	P_i^{max} (kW)	P_i^{min} (kW)	RU_i (kW/h)	RD_i (kW/h)	SUC_i ($)	C_i ($/kWh)
DG 1	600	100	105	120	20	0.040
DG 2	950	200	175	200	30	0.035

and minimum SOE of the ES are equal to 90% and 10% of its energy capacity due to life-time considerations. The initial SOE of ES is 50% of its energy capacity.

It is assumed that only there are three FLs so that they can curtail 20% of their initial consumption in each hour in response to 0.035 $/kWh as an incentive payment. The load-pickup and load-drop rates are considered to be 25 kW/h. Furthermore, two conventional DGs are available with technical characteristics as well as cost terms reported in Table 5.1.

Figure 5.3 represents the trading power of CVPP in the day-ahead market. In this figure, the positive values are assigned to purchased power, while the negative values represent the sold power in the day-ahead market. According to this figure, it is obvious that the CVPP purchases power in time periods with low market prices (early morning and end of the night), whereas it sells power to the market during the day when the market price is relatively high.

The obtained results for the optimal dispatch of various components belonging to CVPP have been reported in Figure 5.4. According to Figure 5.4, when the aggregated generation is more than consumption, the CVPP sells the extra power to the market and vice versa. It is obvious that both DGs are generating so that their generation in the early morning and end of the day is set to minimum, and during the day,

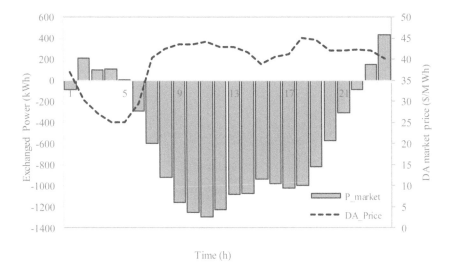

FIGURE 5.3 CVPP trading power vs. expected day-ahead market prices.

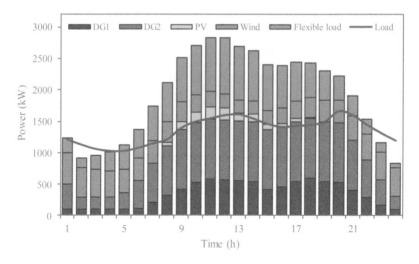

FIGURE 5.4 Power dispatch of components belongs to CVPP.

these DGs generate with full capacity due to the fact that their generation costs are lower than market prices. Moreover, it can be seen that the CVPP utilizes the FLs at all the periods.

The charging and discharging plan of ES during the scheduling horizon has been illustrated in Figure 5.5. As it can be seen, the ES unit charges during the early morning period, particularly at hours 4:00 and 5:00, and injects its power to the grid during the day, specifically between 7:00 and 11:00. The charging and discharging plan of ES unit is completely based on the day-ahead market variations. In fact, in

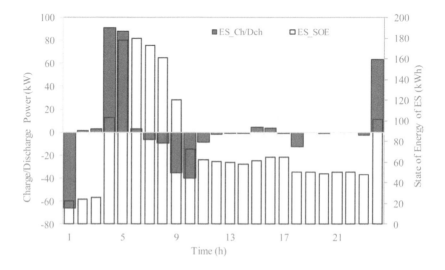

FIGURE 5.5 Charging power, discharging power, and state of energy of ES.

A Comprehensive Smart Energy Management Strategy

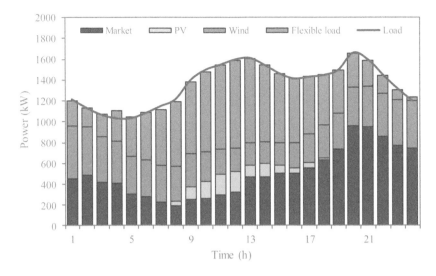

FIGURE 5.6 Power dispatch of components belongs to CVPP with emission cost consideration.

addition to ES role in facilitating the integration of renewable generations such as PV and WT, it uses the energy price arbitrage and makes more profit for CVPP.

The other important points that can affect the CVPP strategy is consideration of pollutant emission cost for DGs due to the fact that the DG's technology is mainly based on diesel burning. In this case, it is assumed that the costs of DGs are multiplied by 1.3. On this basis, both of the DGs never generate and the CVPP is forced to purchase power from market at all time. The power dispatch in this case is shown in Figure 5.6.

As mentioned in the modelling part, the stochastic programming approach has been used here in order to take into account the uncertainties of renewables, load, and market price. In order to evaluate the impacts of forecasted parameters on the power dispatch of various resources, the obtained results for the real case are compared with outputs of the stochastic model. The forecasted scenarios for PV and WT have been compared with the real generation in Figure 5.7.

In such a situation, the obtained results for the cost terms of objective function have been compared in Table 5.2.

TABLE 5.2
Comparison of Cost Terms in Objective Function

Case	Market Transaction Cost ($)	DG Cost ($)	Flexible Load Cost ($)	Total Cost ($)
Scenario-based	−586.5	965.6	420.7	799.8
Real data	−588.9	973.1	414.0	798.2

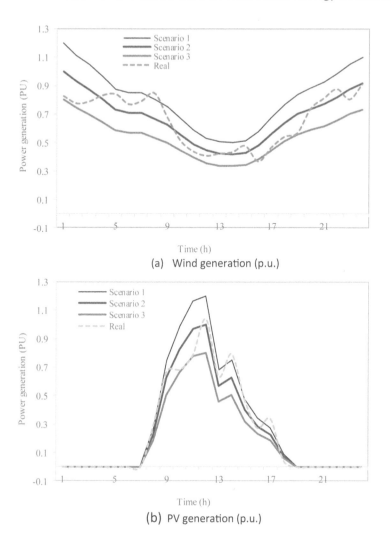

FIGURE 5.7 Scenarios and real data of renewable generation: (a) Wind generation (p.u.) and (b) PV generation (p.u.)

The obtained results for market transactions of the CVPP in two mentioned cases (real and forecasted data) have been compared in Figure 5.8. As observed, the proposed model is relatively robust in the face of forecast tool's error. However, there are some deviations in the number of time slots (red box).

ACKNOWLEDGEMENT

This work has been carried out in the framework of the European Union's Horizon 2020 research and innovation program under grant agreement No 957852 (Virtual Power Plant for Interoperable and Smart isLANDS – VPP4ISLANDS).

A Comprehensive Smart Energy Management Strategy

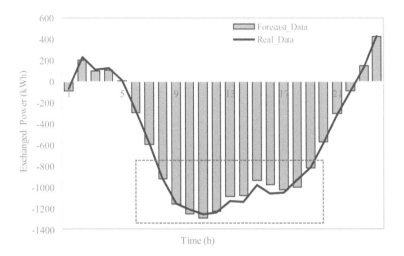

FIGURE 5.8 CVPP trading power comparison in two defined cases.

REFERENCES

1. C. Ramsay, "The virtual power plant: enabling integration of distributed generation and demand," FENIX Bulletin, 2, Jan 2008.
2. S.M.B. Sadati, J. Moshtagh, M. Shafie-khah, J.P. Catalão, "Smart distribution system operational scheduling considering electric vehicle parking lot and demand response programs," Electric Power Systems Research, 160, pp. 404–418, 2018.
3. D.Z. Fitiwi, L. Olmos, M. Rivier, F. de Cuadra, I.J. Pérez-Arriaga, "Finding a representative network losses model for large-scale transmission expansion planning with renewable energy sources," Energy, 101, pp. 343–358, 2016.
4. J.F. Franco, M.J. Rider, M. Lavorato, R. Romero, "A mixed-integer LP model for the reconfiguration of radial electric distribution systems considering distributed generation," Electric Power Systems Research, 97, pp. 51–60, 2013.
5. A.C. Rueda-Medina, J.F. Franco, M.J. Rider, A. Padilha-Feltrin, R. Romero, "A mixed-integer linear programming approach for optimal type, size and allocation of distributed generation in radial distribution systems," Electric Power Systems Research, 97, pp. 133–143, 2013.
6. M. Braun, P. Strauss, "A review on aggregation approaches of controllable distributed energy units in electrical power systems," International Journal of Distributed Energy Resources, 4, 4, pp. 297–319, 2008.

6 Virtual Energy Storage Systems for Virtual Power Plants

Saif S. Sami, Yue Zhou, Meysam Qadrdan, and Jianzhong Wu
School of Engineering, Cardiff University, Cardiff, United Kingdom

CONTENTS

6.1 The Concept of VESS 120
6.2 Components of VESS 121
    6.2.1 Flexible Demand Units 121
    6.2.2 Energy Storage Systems 122
6.3 Enabling Technologies For VESS 126
6.4 Potential Applications of VESS in Power Systems 127
6.5 The VESS As an Integral Part of the VPP 128
6.6 The Potential Benefits of VESS in Power Systems 128
6.7 Control Schemes of VESS 129
6.8 A Frequency Control Scheme of VESS 130
    6.8.1 Modelling of Components of the VESS 130
    6.8.2 The Central Controller and Local Controllers of
           Refrigerators and Flywheel Energy Storage Units 130
    6.8.3 Case Study 133
6.9 A Voltage Control Scheme of VESS 134
    6.9.1 Modelling of Components of the VESS 135
    6.9.2 Distributed Controllers of the Bitumen Tanks and
           Battery Energy Storage System 136
    6.9.3 Case Study 138
6.10 Summary 142
Acknowledgement 142
References 142

NOMENCLATURE

ΔP_{VESS_req} the required power change of the virtual energy storage system (VESS) (MW)

R_{VESS} the droop control value of the VESS

ΔP_{FESS} the power change of flywheel energy storage system (FESS) units (MW)

DOI: 10.1201/9781003257202-7   **119**

f	the grid frequency (Hz)
f'	the modified frequency for controlling FESS units (Hz)
R_{FESS}	the droop control value of FESS units
Δf	the frequency deviation (Hz)
T_{Ca}	the cavity temperature of a refrigerator (°C)
T_{low}, T_{high}	the set-points of the temperature controller of a refrigerator or a bitumen tank (°C)
S_T	the state signal of the temperature controller of a refrigerator or a bitumen tank
F_{ON}, F_{OFF}	the set-points of the frequency controller of a refrigerator (Hz)
S_H, S_L	the state signals of the frequency controller of a refrigerator
ω	the velocity of a FESS unit (rad/s)
$F_{Chrg}, F_{Dischrg}$	the set-points of the coordinated frequency controller of a FESS unit (Hz)
$S_{Chrg}, S_{Dischrg}$	the state signals of the coordinated frequency controller of a FESS unit
$R_{adaptive}$	the adaptive droop control value of a FESS unit
Δf^{max}	the maximum frequency deviation (Hz)
$P_{FESS\_capacity}$	the rated power capacity of a FESS unit (MW)
T_{BT}	the temperature of the bitumen in a bitumen tank (°C)
V	the voltage of a busbar (p.u.)
V_{ON}, V_{OFF}	the set-points of the voltage controller of a bitumen tank (p.u.)
S_{HV}, S_{LV}	the state signals of the voltage controller of a bitumen tank
V_{high}, V_{low}	the set-points of the voltage controller of the BESS (p.u.)
ΔP_{ES}	the change in the active power of the BESS (p.u.)
ΔQ_{ES}	the change in the reactive power of the BESS (p.u.)
M_{i_ES}	the voltage sensitivity factor with regard to active power (voltage p.u./active power p.u.)
N_{i_ES}	the voltage sensitivity factor with regard to reactive power (voltage p.u./reactive power p.u.)
BESSp	the active power output of the BESS (MW)
BESSq	the reactive power output of the BESS (MVAr)

6.1 THE CONCEPT OF VESS

A virtual energy storage system (VESS) is defined as cooperation between different controllable distributed energy resources (DERs), such as flexible demand units and small-capacity energy storage units, to provide efficient power services. A VESS, through virtually sharing DERs' storage potential, functions similarly to a large-capacity conventional energy storage system. Other DERs such as distributed generation (DG) and multi-vector energy resources, e.g. combined heat and power (CHP) systems, can also be included in the VESS, as shown in Figure 6.1. The VESS scheme addresses the uncertainty associated with the response from flexible demand by the coordination with energy storage systems (ESSs). Unlike a virtual power plant (VPP), a VESS coordinates DERs to operate as a single large-capacity ESS, which stores the surplus electricity energy and releases it based on the system requirements.

Virtual Energy Storage Systems for Virtual Power Plants

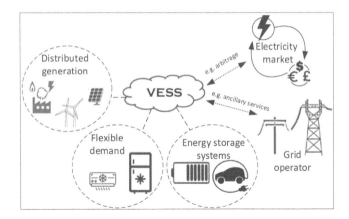

FIGURE 6.1 The concept of a virtual energy storage system (VESS) in smart grids, where a VESS can provide various services within several markets and for different operators.

A VESS is realised through an aggregator, allowing small-capacity flexible demand, ESSs and other DERs to access the wholesale market and provide ancillary services to electricity transmission and distribution networks. A VESS aggregator is an intermediate between a power system operator or service recruiter and a group of VESS components such as flexible demand or ESS owners.

Other definitions for VESS are presented by several researchers recently, referring to a single flexible demand unit, i.e. residential air conditioners [1], or an aggregation of flexible demand units, such as refrigerators [2], air-conditioning loads [3, 4] and electric vehicles [5, 6]. However, without ESS, these systems exhibit a high degree of response uncertainty. In contrast, a successful adoption of the VESS concept, which refers to an aggregation of ESS, PV systems and flexible demand units, was used for the voltage regulation of distributed networks based on dynamic pricing [7].

6.2 COMPONENTS OF VESS

6.2.1 FLEXIBLE DEMAND UNITS

Demand response (DR) is a change in the electricity consumption of loads from their normal consumption patterns. This shift in electricity consumption can be in response to electricity price changes or incentives paid by power system operators to improve system reliability [8].

Based on the amount of electricity consumption, flexible demand can be classified into (1) large industrial, commercial and other non-domestic demands, (2) small industrial, commercial and public demands and (3) residential demand. A large flexible demand often participates directly in DR programmes, while small flexible demand units are commonly aggregated by intermediaries which are called DR providers (DRPs), curtailment service providers (CSPs) or aggregators of retail customers (ARC) [9].

The use of different flexible demands to support power systems is well established. These include industrial demand such as steelworks, public utility demand

122   Virtual Power Plant Solution for Future Smart Energy Communities

such as water supply and wastewater treatment plants, health and educational buildings such as hospitals and universities and commercial demand such as retailers [10]. However, the share of domestic demand in DR programmes has been limited since the associated costs are high per participation capacity. Economic benefits are also considered small to these consumers as many have flat electricity prices [11]. Yet, a study of Great Britain (GB) revealed that the acceptance of DR programmes at the domestic level is expected to reach 80% in 2050, mainly through smart home appliances which can shift approximately 11% of their peak demand through time-of-use programmes [12].

DR programmes are commonly classified into price-based and incentive-based programmes based on how flexible demand is recruited. A comparison of different programmes is depicted in Table 6.1 [8, 13]. In the price-based DR programmes, consumers adjust their electricity consumption to changes in the electricity price. While incentive-based DR programmes are implemented through interruptible or curtailment contracts, in which consumers are paid to reduce or shift their electricity consumption. These DR programmes are deployed at different time scales within the power system, as shown in Figure 6.2. Sustaining the power system security and reliability is the main motivation for establishing incentive-based programmes, while economic aspects drive the price-based programmes. Typically, DR in incentive-based programmes is activated by events such as large frequency deviations, while changes in electricity prices often trigger DR in price-based programmes. Currently, flexible demand provides different services in several European countries and the USA, which include frequency regulation, congestion management and voltage regulation to a transmission system operator (TSO) at the transmission system level and congestion management and voltage regulation to a distribution system operator (DSO) at the distribution network level [14].

For example, in the USA, the potential flexible DR through incentive-based DR programmes at the system peak hour exceeded 30 GW in 2018, which consists of approximately 15 GW from the industrial, 7 GW from the commercial and 8 GW from the residential flexible demand. Additionally, the participation of DR in a regional transmission organization (RTO) or an independent system operator (ISO) wholesale DR programmes in 2019 exceeded 32 GW, which was approximately 6.6% of peak demand [15].

6.2.2 ENERGY STORAGE SYSTEMS

ESSs are typically classified based on the stored energy form into electrical, mechanical, electrochemical and thermal ESSs, as shown in Figure 6.3 [16] and Table 6.2 [10]. ESSs can also be categorised based on the power and energy densities or ratings into high-energy density ESSs and high-power density ESSs. The high-energy density ESSs include pumped hydro ESS, compressed-air ESS, thermal ESS and hydrogen-based ESS, which are typically used for energy management applications. The high-power density ESSs include supercapacitor ESS, flywheel ESS and superconducting magnetic ESS, which are often used for power management applications. Many types of battery ESS have high energy and power densities, and hence, they might be suitable for both power and energy management applications.

Virtual Energy Storage Systems for Virtual Power Plants

TABLE 6.1
Demand Response Programmes [8, 13]

Programme Name		Description	+Advantage/–Disadvantage
Price-based	Time-of-use	This programme uses different electricity prices for different time blocks, typically predefined for a 24-hour day.	– A price scheme is offered to all customers with different consumption levels.
	Real-time pricing	This programme often provides an hourly fluctuating electricity price to reflect changes in the wholesale electricity price.	– Customers should respond rapidly to prices changes.
	Critical peak pricing	In this programme, the electricity price is based on the time-of-use programme. However, certain defined conditions trigger much higher event prices that replace the normal prices.	
Incentive-based	Direct load control	In this programme, on short notice, the programme operator remotely controls customers' electrical appliances.	– Customers must grant the operator a level of authority to shift or curtail certain loads.
	Interruptible/ curtailable service	In this programme, consumers agree to reduce load during system contingencies. The curtailment options are integrated into retail tariffs.	– Failing to curtail contracted loads leads to penalties.
	Demand bidding/ buyback	In this programme, consumers offer bids for load curtailment based on wholesale electricity market prices.	
	Emergency demand response	In this programme, consumers are offered incentive payments to reduce their loads during periods when the system is short of reserve.	+ Consumers incur a credit or electricity price discount.
	Capacity market	In this programme, consumers are offered incentive payments to reduce their loads as a replacement to conventional generation. Customers normally receive intra-day notice of events.	+ Incentives include upfront payments.
	Ancillary services market	In this programme, consumers bid for load increase or curtailments. If their bids are accepted, they are paid for committing to be on standby and when their load curtailments are required.	

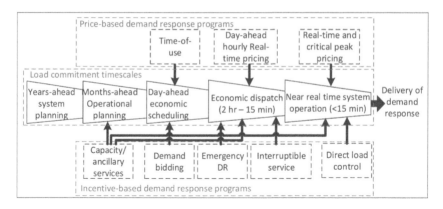

FIGURE 6.2 The deployment of demand response in power systems [8].

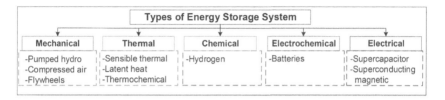

FIGURE 6.3 Energy storage systems used in the power system [17].

TABLE 6.2
Types of Energy Storage Systems for Power Systems [10]

Type	Description	+ Advantages/− Disadvantages
Pumped hydro energy storage (PHES)	The energy is stored by pumping water from the lower reservoir to the upper reservoir and deployed when water is discharged in the opposite direction through the turbine. The energy stored depends on the height difference between the reservoirs.	+ Zero emissions produced, a long life, a large storage scale, a fast reaction and low maintenance. − The geographical constraint and the capital costs.
Compressed air energy storage system (CAES)	The system compresses air, i.e. the medium of storage, normally in underground salt caverns to store energy. The energy is regained when the stored air is decompressed and heated by a combustion gas turbine.	+ Can be deployed at a large scale. − The necessity of a fuel source and the geographical constraint.
Flywheel energy storage system (FESS)	The system consists of a large inertia flywheel coupled with an electrical machine. The conversion of electrical to mechanical energy occurs through accelerating the velocity of the flywheel and retrieved by decelerating the velocity.	+ Has a large number of charging and discharging cycles, i.e. a long life, and a fast response. − Has a high self-discharge.

(Continued)

Virtual Energy Storage Systems for Virtual Power Plants

TABLE 6.2 *(Continued)*
Types of Energy Storage Systems for Power Systems [10]

Type	Description	+ Advantages/− Disadvantages
Thermal energy storage system (TESS)	The energy is stored in the forms of cold, heat or their combination. The system is classified based on physical principles into A-latent heat storage, based on the phase change materials; B-sensible thermal energy storage, based on the temperature difference; C-thermochemical energy storage.	+ Inexpensive and has no geographical constraints. − Has a moderate efficiency.
Hydrogen-based energy storage system (HESS)	This system essentially consists of an electrolyser and fuel cells. Fuel cells generate electricity by composing hydrogen and oxygen into water, while electrolyser consumes electricity to decompose water.	+ Has a separate process for the generation, storage and deployment. − Has a low overall efficiency.
Lithium-ion battery energy storage system (Li-ion BESS)	The typical structure of a lithium-ion battery consists of a cathode made of lithium metal oxide, a graphite anode and an electrolyte consisting of a solution of a lithium salt in a mixed organic solvent embedded in a separator felt.	+ The system has a high power, high permissible depth of discharge. − Has a relatively high cost. Yet, the cost is declining rapidly in recent years.
Lead-acid battery energy storage system (Lead-acid BESS)	The system is made of stacked cells, immersed in an electrolyte of a dilute solution of sulphuric acid (H_2SO_4). The negative electrode of each cell is sponge lead (Pb), while the positive electrode is composed of lead dioxide (PbO_2).	+ This is the least expensive BESS. − Has a low permissible depth of discharge and a limited cycling capability.
Sodium–sulphur battery energy storage system (Na-S BESS)	The system operates at a temperature of about 300°C to keep the electrode materials in a molten state, hence reducing resistance to the sodium ions flow through the β-alumina solid (β-Al_2O_3) electrolyte.	+ Has a high efficiency and a low maintenance requirement. − Safety and thermal management are the key disadvantages.
Nickel battery energy storage system (Ni BESS)	The main system active materials of the positive and negative electrodes are nickel (Ni) and cadmium (Cd), respectively. Aqueous alkali solution is acting as the electrolyte.	− Has considerable costs, toxicity and the memory effect problem.
Flow battery energy storage system (Flow BESS)	The system consists of an electrolyte that contains one or more dissolved electroactive materials flowing through a power cell/reactor (i.e. where the chemical energy is converted to electricity). The three major types are vanadium redox battery (VRB), zinc–bromine battery (ZBB) and polysulphide bromide battery (PSB).	+ The ability to deliver long-term charging/discharging and a negligible self-discharge. − Has a moderate efficiency.

(Continued)

TABLE 6.2 *(Continued)*
Types of Energy Storage Systems for Power Systems [10]

Type	Description	+ Advantages/– Disadvantages
Super- or ultra-capacitors energy storage system	The system is based on two-conductor electrodes along with an insulator (electrolyte and a porous membrane). The energy stored is directly proportional to the capacitance capacity and the square of the voltage between its terminals.	+ Has a long life, no discharge depth limitations and relatively high-power capacity. – Has a small energy capacity and a high cost.
Superconducting magnetic energy storage (SMES)	The energy is stored in the magnetic field generated by the direct current flowing through a coiled wire at cryogenic temperature. The stored energy is proportional to the coil self-inductance and the square of the current passing through it.	+ Has a high efficiency and a rapid response, albeit for short periods of time. – Requires continuous energy to cool the coil, environmental issues related to the strong magnetic field.

ESSs can provide a wide range of services to bulk power systems and microgrids. For example, these services can improve the power quality and system stability and support the integration of renewable generation, as shown in Figure 6.4 [18].

6.3 ENABLING TECHNOLOGIES FOR VESS

The conceptual framework of the VESS involves the use of various smart technologies to facilitate VESS components' coordination and support the efficient monitoring and control. These enabling technologies are described as follows:

- **Smart switches**
  Smart switches are used to remotely control particular end-user loads, such as a refrigerator, or a [15] heating, ventilation and air conditioning (HVAC) system (in this particular case, it might be also known as smart thermostats). This smart thermostat can regulate room temperatures according to remote signals sent from the master controller of the VESS.

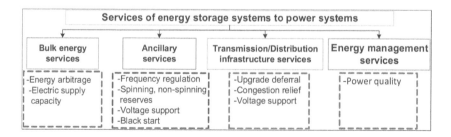

FIGURE 6.4 Potential applications of energy storage systems in power systems.

Virtual Energy Storage Systems for Virtual Power Plants 127

- **Smart building management technologies**
  These technologies enable an agent to aggregate all flexible loads within a building to act as a VESS component. The flexible loads can be dishwashers, washing machines, dryers, swimming pool pumps, EV charging stations, behind-the-meter ESSs, elevators and some types of lights. Through two-way communication systems, the agents receive control signals from the master controller of the VESS and report back all the information on available flexible demand considering customers' preferences and comfort levels.

- **Communication technologies**
  Communication systems are a vital part of implementing VESS control schemes. One-way communication systems are simple to implement and more cost-effective. However, an accurate verification of the response of VESS components is difficult to achieve. Two-way communication systems are more expensive, yet monitoring and identifying the available level of flexible loads and ESSs in near real time. These communication systems can include home area networks (HANs), neighbourhood area networks (NANs) and wide-area networks (WAN). Both wired and wireless communication technologies can be used to link different VESS components. Wired communication technologies include narrowband and broadband power line communications (PLCs) and fibre optics, while Zigbee (IEEE 802.15.4) and Wi-Fi (IEEE 802.15.4) are typical wireless communication technologies that can be adopted.

6.4 POTENTIAL APPLICATIONS OF VESS IN POWER SYSTEMS

Recent developments in integrated circuits, such as inexpensive smart switches and information and communications technologies, improve the advanced monitoring and control functionalities. These developments accelerate the realisation of the VESS concept. The VESS forms a synthetic ESS at distribution and transmission levels, and hence it can provide services at both levels. The VESS potential applications are presented as follows.

- **Facilitate the integration of DG in distribution networks**
  A VESS can smooth the variations in the power output of intermittent renewable generation. This is beneficial to increase the hosting capacity of DG in distribution networks. The hosting capacity is the total DG capacity allowed into the network without violating network constraints, e.g. voltage and thermal constraints, under the minimum loading condition. A VESS will address variations in the DG power output to restrict the voltage deviations and power flow below the limits.

- **Reduce reserve margins**
  A VESS can reduce the required spinning reserve capacity and increase the generators' loading level, since the available VESS capacity can be continuously reported to the system operator as a fast-acting spinning reserve.

128   Virtual Power Plant Solution for Future Smart Energy Communities

- **Defer transmission and distribution systems reinforcement**
  Transmission and distribution systems are often sized to accommodate the expected peak demand. Therefore, reducing peak demand, through the VESS, allows the system reinforcement to be deferred. In addition, the VESS can increase the utilisation of transmission and distribution networks by providing immediate actions to avoid potential network congestions, and hence the transmission and distribution systems upgrade can be postponed.
- **Provide other ancillary services**
  As a result of aggregation, the VESS can provide different ancillary services to the power system operator, such as frequency response services, since it provides a faster response and higher ramp rates than the conventional generation units.

6.5 THE VESS AS AN INTEGRAL PART OF THE VPP

Through the aggregation of various types of DERs, the VESS poses the characteristics of an ESS with high power and energy ratings. A VESS facilitates required flexibility for the VPP to shift electrical energy from one period of time to another, hence extending the VPP potential over widespread power system applications. A VESS can improve the reliability and quality of the power supply provided by the VPP. The VESS involvement increases the performance of the VPP and extends its capabilities that include, but are not limited to, the following:

- **Mitigate the fluctuation of renewable generation**
  A VESS could smooth the fluctuation of power generated by renewable generation within the VPP. A VESS deployment across a significant part of the VPP geographical area can eliminate many DG curtailments due to network constraints' violations. A VESS can alleviate renewable generation forecasting errors enabling an efficient and accurate service delivery by the VPP.
- **Enhance the quality of supply**
  The future VPP is anticipated to cope with essential supporting or balancing services, e.g. congestion management, black start capability and inertial response capability [19]. A VESS will assure that the VPP affirms these obligations. The future VPP is expected also to provide an uninterrupted power supply where the VESS retains the consistency in VPP operations. Additionally, the provided electrical service by the VPP will be disruption-free of voltage swells, sags and spikes. A VESS can also support the VPP to prevent blackouts and enhance ancillary services provided by the VPP [20].

6.6 THE POTENTIAL BENEFITS OF VESS IN POWER SYSTEMS

The benefits of the VESS may vary based on several factors such as controllers' design and performance, the targeted market as well as enabling technologies utilised and the VESS components involved. These benefits at different levels of the power system can be categorised as follows:

Virtual Energy Storage Systems for Virtual Power Plants 129

- **Benefits for VESS components owners**
  The components owners can obtain additional revenue through being a part of VESS, which arbitrages in the wholesale electricity market and/or provides ancillary services to power system operators. In addition, the advanced communication and control functionalities of VESS can help improve consumers' economic and environmental awareness for electricity consumption, which may result in better electricity-saving behaviours and bring diversified options for electricity costs and emission management.
- **Benefits for transmission and distribution systems**
  The VESS can reduce the overall power system losses, alleviate some of systems constraints and improve systems reliability. The VESS can defer the reinforcement of electrical networks without comprising network-assigned reliability levels.
- **Benefits for supply side of power systems**
  The VESS can reduce required generation during peak times, and hence investment in peaking units can be reduced. Additionally, as the VESS allows more renewable energy generation into the energy market, the influence of conventional generation companies on market prices may be reduced.
- **Benefits for VPP**
  The VESS can boost the capacity of VPP in participating in the energy market, potentially transforming a price-taker VPP into a large-scale generation entity and a price-maker [21]. Consequently, the VPP has a high influence in the electricity market that can adjust market prices to earn higher profits. The VESS will also increase the capability of VPP to provide ancillary services to power systems.

6.7 CONTROL SCHEMES OF VESS

The control scheme of a VESS can be central or distributed, depending mainly on the required information exchange among VESS components, spatial distribution of these components and the allocated budget involved. Central control schemes use sophisticated two-way communication systems at high-time resolutions that allow the VESS to furnish ancillary services such as frequency and voltage support and spinning reserve. For example, a VESS-centralised frequency controller can send a turning ON/OFF signal to home appliances, such as heat pumps, water heaters or refrigerators, after a frequency rise or dip event. Also, the near real-time VESS availability enables the accommodation of more distributed renewable generation that otherwise would be curtailed. The centralised controller can be managed by a VESS aggregator. The computation burden of the centralised control scheme continues to be a barrier, since numerous variables of VESS components need to be considered.

To address the problems associated with two-way communication systems, such as latency, packet loss and costs, decentralised controllers were investigated. For instance, a decentralised VESS frequency controller can manipulate the temperature set-points of air conditioners or refrigerators to vary in-line with the frequency

130   Virtual Power Plant Solution for Future Smart Energy Communities

deviations, hence altering their power consumption in response to the frequency deviations.

A hierarchical control scheme can include central and local controllers and utilise an amalgam of communication technologies to realise. A deep analysis of control requirements and capabilities is required to tailor the right set of these technologies.

Two applications were chosen to represent the potentials of the VESS in Sections 6.8 and 6.9. In the first application, the VESS is providing balancing services, i.e. frequency response, to the TSO. While in the second application, the VESS is supporting the voltage control in a distribution network with a high penetration of distributed renewable energy resources, hence assisting the DNO.

6.8 A FREQUENCY CONTROL SCHEME OF VESS

The frequency control scheme of a VESS was developed. The VESS can provide low, high and continuous frequency responses. In this application, the VESS coordinates domestic refrigerators and flywheel energy storage units to deliver a certain amount of frequency response at a lower cost compared with an equal capacity of using units of flywheel energy storage only [17].

6.8.1 MODELLING OF COMPONENTS OF THE VESS

The thermodynamic model of refrigerators, which was adopted from [22], is utilised in this study. The model uses two first-order differential equations to relate the rate of change in the cavity and evaporator temperatures with time to parameters that describe the thermal characteristics of a refrigerator.

A simplified model of a FESS, which was developed in [23], is used in this study. The FESS is essentially an electrical machine coupled with a high inertia flywheel and is connected to the grid through back-to-back converters.

6.8.2 THE CENTRAL CONTROLLER AND LOCAL CONTROLLERS OF REFRIGERATORS AND FLYWHEEL ENERGY STORAGE UNITS

Following a frequency deviation (Δf (Hz) in Figure 6.5), the required frequency response of the VESS (ΔP_{VESS_req} (MW) in Figure 6.5) is determined by the droop control with the droop coefficient (R_{VESS}) of the value of 1%. Hence, a 1% change in grid frequency would trigger a 100% change in the VESS power. First, local controllers of refrigerators respond to the frequency deviation. Then, FESS units eliminate the power mismatch between the change in refrigerators' power consumption and the power required from the VESS. Consequently, FESS units compensate for the uncertainty in the response of refrigerators. The required power from FESS units ($\sum \Delta l_{FESS\text{-}req}$ (MW) in Figure 6.5) is decided by a modified frequency value f' (Hz) through the droop setting (R_{FESS}). The local controllers of FESS units respond to the modified frequency. It is assumed that a fast two-way communication system is available for receiving the power of refrigerators and sending the modified frequency to FESS units.

Virtual Energy Storage Systems for Virtual Power Plants

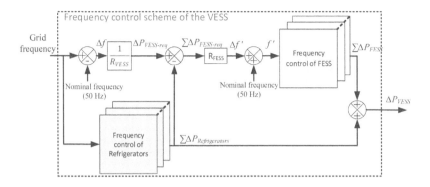

FIGURE 6.5 The central frequency controller of the VESS.

A local frequency controller is integrated, through a lookup table, to the internal temperature controller of a refrigerator, as shown in Figure 6.6 and Table 6.3. The temperature controller continuously measures the cavity temperature (T_{Ca}) (°C) and compares it with set-points (T_{low} and T_{high}) (°C). If the cavity temperature reaches T_{high}, the controller generates state signal (S_T) of 1, or it reaches T_{low}, the state signal generated (S_T) is 0. The frequency controller, based on T_{ca}, defines a pair of frequency set-points (i.e. F_{ON} and F_{OFF} (Hz)), which dynamically varies with the temperature T_{Ca}. It compares the measured frequency (f) to these set-points to determine the state signals (S_H and S_L). The range of F_{ON} is 50–50.5 Hz and the range of F_{OFF} is 49.5–50 Hz, which are consistent with the steady-state limits of grid frequency in the GB power system.

In the case of a population of refrigerators, a refrigerator having a lower temperature than others will have a higher F_{ON} and a higher F_{OFF} values as indicated in Figure 6.6. If f drops, refrigerators will start switching OFF from the refrigerator with the lowest T_{ca}, because it will take the longest time to reach the high-temperature limit. In contrast, refrigerators will start switching ON from the refrigerator with the highest T_{ca} when f rises above the nominal frequency value. The higher the frequency variation is, the larger number of refrigerators will be committed to respond. When a temperature-diversified population of refrigerators

TABLE 6.3
Lookup Table in Figure 6.6

Row	S_T	S_L	S_H	S_{final}
1	0	0	0	0
2	0	0	1	1
3	0	1	0	0
4	1	0	0	1
5	1	0	1	1
6	1	1	0	0

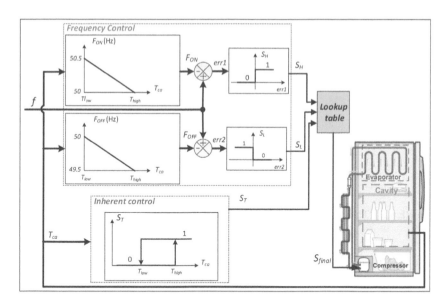

FIGURE 6.6 The integrated control of the refrigerator [17].

is considered, the number of refrigerators committed increases linearly with the increase in frequency variations.

The local frequency controller of the FESS consists of the coordinated control and the adaptive droop control as depicted in Figure 6.7. The coordinated control determines which unit to commit, while the adaptive droop control regulates the power output of the committed units. The coordinated control is similar to the local frequency control of refrigerators presented early. However, the temperature is replaced by the velocity (ω) (rad/sec) of the FESS unit, which also represents the state of charge (SoC) of the unit. The coordinated control, based on ω, defines a pair of frequency set-points (F_{Chrg} and $F_{Dischrg}$) and compares the grid frequency to these set-points to determine the state signals (S_{Chrg} and $S_{Dischrg}$). The coordinated control, through the OR logic gate and a switch shown in Figure 6.7, ensures that the number of FESS units committed is linearly increasing with the increase in frequency deviations.

The adaptive droop control value ($R_{adaptive}$) is inversely proportional to frequency deviations (Δf) as shown in Equations (6.1.a) and (6.1.b). Dictated by coordinated control, a small frequency deviation (Δf) triggers only a small number of FESS units to commit. Therefore, a droop value $R_{adaptive}$ greater than the conventional droop value R_{FESS} is required to increase the change of power output. When the frequency deviation increases and reaches the frequency deviation limits (Δf^{max}) (±0.5 Hz in the GB power system), all FESS units will be triggered to commit and $R_{adaptive}$ equals R_{FESS}. R_{FESS} is set to 1%, which indicates that a FESS unit will provide 100% power output change if frequency deviation is equal to or higher than 1% of the nominal frequency value.

$$R_{FESS} = \frac{\Delta f^{max}}{P_{FESS\_capacity}} \qquad (6.1.a)$$

Virtual Energy Storage Systems for Virtual Power Plants

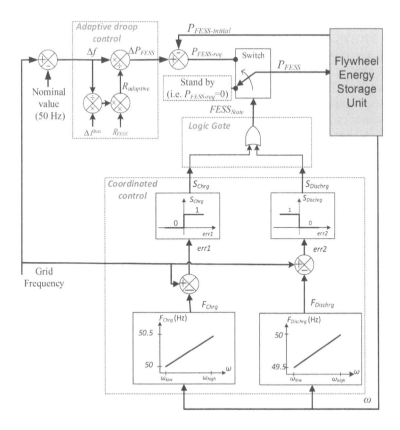

FIGURE 6.7 The frequency control of the FESS.

$$R_{adaptive} = \frac{\Delta f^{max}}{\Delta f} \times R_{FESS} \quad (6.1.b)$$

where $P_{FESS\_capacity}$ (MW) is the rated power capacity of the FESS unit.

6.8.3 Case Study

To assess the performance of the VESS to provide low-frequency response services, it is connected to the simplified GB power system model adopted from [24, 25]. Further details can be found in [10]. Three scenarios were compared:

Scenario 1: No FESS/VESS.
Scenario 2: Only FESS (60 MW of FESS is used).
Scenario 3: VESS (a large number of refrigerators with the power reduction potential of 40–60 MW and 20 MW of FESS are used).

The availability of refrigerators to be switched OFF varies over the day from 13.2% to 18.5% [26]. Considering the participation of 3,220,000 refrigerators (0.1 kW for each), a maximum power reduction of 60 MW and a minimum power reduction of 40 MW is expected. It is worth noting that the availability of refrigerators to be switched ON over the day is 50%–56% [26], and hence, the power that can be increased ranges from 160 MW to 180 MW. This reveals that refrigerators have more potential to provide a response to the frequency rise than to the frequency drop. Each FESS unit has a power capacity of 50 kW and an energy capacity of 30 kWh, similar to the commercial FESS in [27]. Hence, the 20 MW of FESS consists of 400 units (**Scenario 3**) and 60 MW of FESS consists of 1200 units (**Scenario 2**).

Simulations were carried out by applying a loss of a generation of 1.8 GW to the GB power system. This case simulates the discharging phase of the VESS. Results are depicted in Figures 6.8 and 6.9. The frequency drop in Figure 6.9 is restricted by 60 MW of response (please see Figure 6.9a) from either FESS (Scenario 2) or VESS (Scenario 3). Since 60 MW of response is small in a 20-GW system, the frequency improvement is only 0.01 Hz, which is hardly noticeable. The capacity of FESS in the VESS (Scenario 3) is only one-third of that in Scenario 2, but VESS provided a similar amount of frequency response to that of FESS in Scenario 2. The reduced capacity of FESS in Scenario 3 will reduce the cost significantly compared to Scenario 2. An economic evaluation of the benefits of VESS for the provision of frequency response services (low, high and continuous responses) is presented in [17].

6.9 A VOLTAGE CONTROL SCHEME OF VESS

A voltage control scheme of the VESS was developed to support the distribution network voltage and hence allows more renewable generation in the distribution

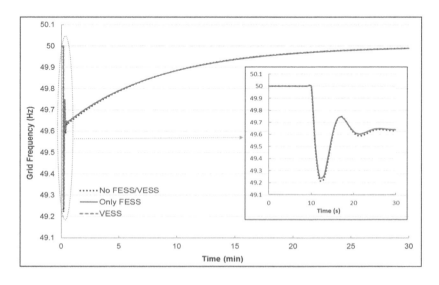

FIGURE 6.8 Variation of grid frequency after the loss of generation [17].

Virtual Energy Storage Systems for Virtual Power Plants 135

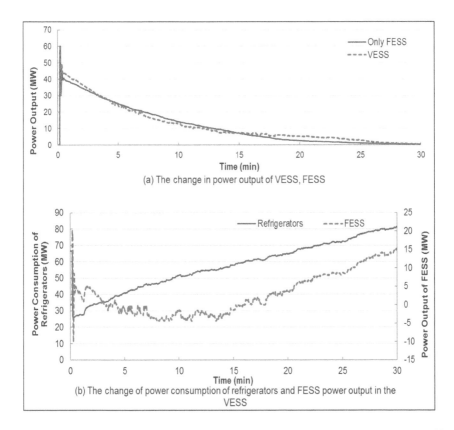

FIGURE 6.9 The change in power output/consumption of (a) the VESS against the FESS and (b) refrigerators and the FESS within the VESS [17].

network. Modelling of VESS components is presented, and then their voltage controllers are presented. In this application, the VESS is an aggregation of industrial bitumen tanks (BTs) and the BESS.

6.9.1 Modelling of Components of the VESS

A thermodynamic model of a BT adopted from [24] is used. The model depicts temperature variations of BT with time, which is captured by a single first-order differential equation. Each bitumen tank has an internal temperature controller which keeps the stored bitumen within a range of temperatures, i.e., T_{low} and T_{high} (°C).

A simplified model of a BESS developed in [23] is used. The model consists of a generic battery model and a simplified power electronics model.

6.9.2 Distributed Controllers of the Bitumen Tanks and Battery Energy Storage System

A local voltage controller was added to each BT's internal temperature controller [28], which has a similar structure to the frequency controller of refrigerators presented in Section 6.8.2. The voltage controller alters BT's power consumption based on local voltage measurements as shown in Figure 6.10. The temperature controller measures the bitumen temperature (T_{BT}) and generates state signals (S_T). The voltage controller measures the voltage (V) in (p.u.) of the busbar connecting BT and defines a pair of voltage set-points (V_{ON} and V_{OFF}) (p.u.) and compares V to these set-points to determine the state signals (S_{HV} and S_{LV}). The final switching signal (S_{final}) to the heater is then determined by a lookup table (see Table 6.3), which ensures the priority of the temperature control. Therefore, the extra voltage control will not undermine the thermal storage function of BTs.

The voltage controller of the BESS monitors, e.g. through remote terminal units (RTU), the most vulnerable busbars with respect to the voltage violation. These busbars are often the ones loaded heavily or connecting a large DG capacity. The controller initially selects, through the selection algorithm, the designated busbar (i), as

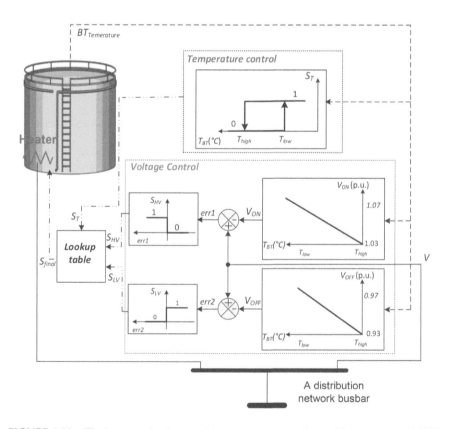

FIGURE 6.10 The integrated voltage and temperature controllers of the bitumen tank [28].

Virtual Energy Storage Systems for Virtual Power Plants

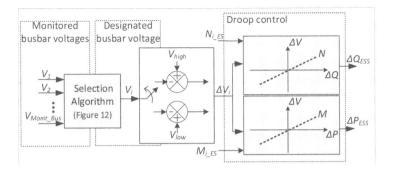

FIGURE 6.11 The voltage controller of the battery energy storage system.

shown in Figure 6.11. The high and the low voltage limits (i.e. V_{high} and V_{low}) were set to 1.06 and 0.94 p.u., respectively, based on [29]. The required changes in the active (ΔP_{ES} in p.u.) and the reactive powers (ΔQ_{ES} in p.u.) of the BESS were calculated using (6.2).

$$\Delta V_i = M_{i\_ES} \times \Delta P_{ES} + N_{i\_ES} \times \Delta Q_{ES} \quad (6.2)$$

where M_{i_ES} is the voltage sensitivity factor (in voltage p.u./active power p.u.), which relates the change in the active power of the BESS to the change in the voltage of busbar i and N_{i_ES} is the voltage sensitivity factor (in voltage p.u./reactive power p.u.), which relates the change in the reactive power of the BESS to the change in the voltage of busbar i. The calculation of voltage sensitivity factors is presented in [10].

The rule-based selection algorithm, depicted in Figure 6.12, is implemented as follows:

1. If all monitored busbar voltages are within limits, no designated busbar is assigned and the BESS takes no action.
2. Among all monitored busbars, select the two busbars with the largest voltage deviations. Since in the worst case, the network will suffer from a high voltage problem at a busbar and a low voltage at another.
3. If both selected busbar voltages violate limits in opposite directions, i.e. one above the high limit and the other is below the low limit, no designated busbar is assigned as well since reducing the voltage violation at one busbar may lead to a more severe voltage violation at the other busbar.
4. If both busbars have a similar direction of voltage deviations, i.e. voltages of both busbars are above/below a nominal value, the designated busbar is the busbar with the higher voltage violation. The BESS will then charge/discharge to bring the voltage of the designated busbar back to the voltage limit.
5. If the two busbars have opposite directions of voltage deviation, the designated busbar is the one not violating voltage limits. The BESS will charge/discharge to push the voltage of the designated busbar to the voltage limit, therefore reducing the other busbars have opposite direct

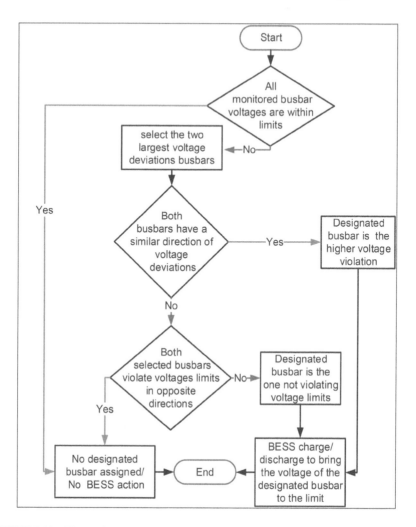

FIGURE 6.12 The designated busbar selection algorithm of the BESS voltage controller.

6.9.3 Case Study

The performance of the proposed VESS voltage control scheme was evaluated using a simplified medium-voltage network, shown in Figure 6.13 [30]. The network supplies a peak load of 38.94 MVA. The network hosting capacity, defined as the total DG capacity under the minimum loading condition, was estimated in [28]. The network hosting capacity is 48.45 MW, which consists of 41.9 MW of wind farms and 6.55 MW of PV systems. The network voltage was only controlled by an on-load tap changing transformer (OLTC) and a voltage regulator (VR) transformer. The network data, half-hourly DG generation and load profiles, were obtained from [30].

For all the network load busbars, except the main busbar (busbar no.1), 30% of loads were assumed to be flexible, i.e. replaced by 9.8 MW, equivalent capacity of

Virtual Energy Storage Systems for Virtual Power Plants 139

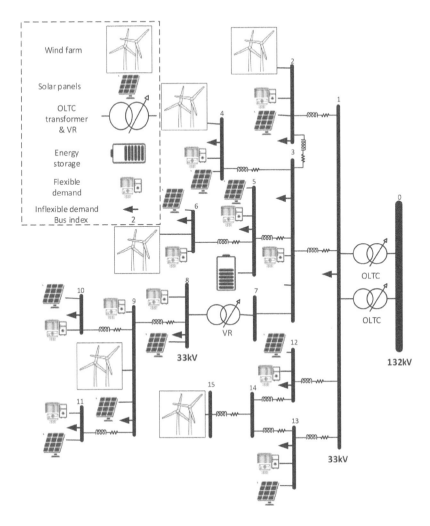

FIGURE 6.13 The distribution network used in the case study [28].

BTs. In addition to BTs, the VESS includes a 2.3-MW/1.4-MWh BESS at busbar no 5, which was installed to compensate for the flexible DR uncertainty. With the presence of flexible demand, the network hosting capacity for DG increased to 60.25 MW.

The VESS control scheme accounts for OLTC and VR voltage controllers to eliminate any controllers' conflicts or hunting between them. The time delay constraints for VESS elements and network transformers are detailed in Table 6.4.

The power flow analysis of 1-minute resolution was carried out to evaluate the proposed VESS control scheme performance over one spring day with high DG power output and low network demand. Results were compared with the base case in which no VESS was used. The half-hourly wind and solar generation profiles and total load of the base case are shown in Figure 6.14. The coincidence of high DG output and low demand led to voltage violations at several busbars in the first five hours of the

TABLE 6.4
VESS and Transformers Control Time Delay

Parameter	Time Delay (min)
τ_{DR}	1
τ_{ESS}	2
τ_{VR}	3
τ_{OLTC}	4

day. Figure 6.15 shows the response of the VESS, where BTs and mostly the reactive power of BESS were sufficient to mitigate the over-voltage caused by high DG. Figure 6.16 depicts the distribution of busbars voltages over the day, with the number of samples being 720 (i.e. 15 busbars over 48 time intervals). The proposed VESS control scheme reduces the number of actions of the OLTC and VR transformers by approximately 30% compared with the base case where no VESS was used [28], hence reducing their maintenance requirements and prolonging their lifespan.

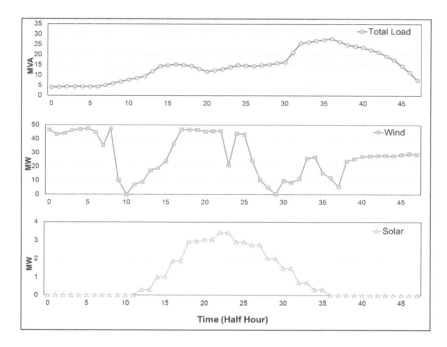

FIGURE 6.14 Wind and solar generation profiles and total load of the base case where no VESS exists in one spring day [28].

Virtual Energy Storage Systems for Virtual Power Plants 141

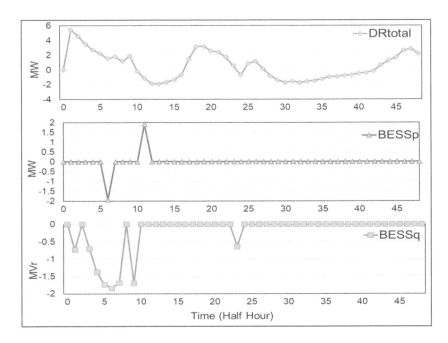

FIGURE 6.15 The response from different VESS components, where DRtotal is the aggregated response power from BTs and BESSp and BESSq are the active and reactive power outputs of BESS [28].

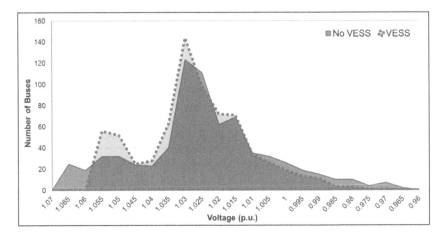

FIGURE 6.16 The distribution of voltages of all busbars over the day [28].

6.10 SUMMARY

In this chapter, a smart energy management paradigm, called a VESS, was presented. A VESS aggregates flexible demand units and small-capacity energy storage units into a single entity which functions similarly to a large-capacity conventional ESS.

An overview of components of a VESS was first introduced. DR programmes that utilise flexible demand in the power system were briefly reviewed. ESSs were classified. Some of enabling technologies for the VESS were briefly listed. Potential applications of VESS in power systems and its potential benefits were discussed.

A frequency control scheme of the VESS was designed. The control scheme provides low, high and continuous frequency response services to the TSO. The centralised control scheme coordinates a large number of domestic refrigerators and a small number of flywheel energy storage units.

A voltage control scheme of the VESS was also designed. This control scheme facilitates a larger penetration of distributed renewable energy generation by supporting the voltage control of distribution networks. The control scheme coordinates bitumen tanks and battery energy storage units by implementing different time delay settings for the voltage controllers.

ACKNOWLEDGEMENT

This work has been carried out in the framework of the European Union's Horizon 2020 research and innovation program under grant agreement No 957852 (Virtual Power Plant for Interoperable and Smart isLANDS – VPP4ISLANDS).

REFERENCES

1. D. Wang *et al.*, "Smart coordination of virtual energy storage systems for distribution network management," *International Journal of Electrical Power & Energy Systems*, vol. 129, p. 106816, 2021.
2. K. N. Kumar, K. Vijayakumar, and C. Kalpesh, "Virtual energy storage capacity estimation using ANN-based kWh modelling of refrigerators," *IET Smart Grid*, vol. 1, no. 2, pp. 31–39, 2018.
3. K. Xie, H. Hui, and Y. Ding, "Review of modeling and control strategy of thermostatically controlled loads for virtual energy storage system," *Protection and Control of Modern Power Systems*, vol. 4, no. 1, pp. 1–13, 2019.
4. K. Meng, Z. Y. Dong, Z. Xu, Y. Zheng, and D. J. Hill, "Coordinated dispatch of virtual energy storage systems in smart distribution networks for loading management," *IEEE Transactions on Systems, Man, and Cybernetics: Systems*, vol. 49, no. 4, pp. 776–786, 2017.
5. A. M. Jenkins, C. Patsios, P. Taylor, O. Olabisi, N. Wade, and P. Blythe, "Creating virtual energy storage systems from aggregated smart charging electric vehicles," *CIRED-Open Access Proceedings Journal*, vol. 2017, no. 1, pp. 1664–1668, 2017.
6. X. Zhu, J. Yang, Y. Liu, C. Liu, B. Miao, and L. Chen, "Optimal scheduling method for a regional integrated energy system considering joint virtual energy storage," *IEEE Access*, vol. 7, pp. 138260–138272, 2019.
7. W. Kang, M. Chen, W. Lai, and Y. Luo, "Distributed real-time power management for virtual energy storage systems using dynamic price," *Energy*, vol. 216, p. 119069, 2021.

8. Q. Qdr, "Benefits of demand response in electricity markets and recommendations for achieving them: A report to the United States Congress Pursuant to Section 1252 of the Energy Policy Act of 2005," US Department of Energy, Washington, DC, USA, Technical Report, February 2006.
9. P. Siano, "Demand response and smart grids—A survey," *Renewable and Sustainable Energy Reviews*, vol. 30, pp. 461–478, 2014.
10. S. S. Sami, "Virtual energy storage for frequency and voltage control," *Ph.D. thesis*, Cardiff University, 2017.
11. K. T. Ponds, A. Arefi, A. Sayigh, and G. Ledwich, "Aggregator of demand response for renewable integration and customer engagement: Strengths, weaknesses, opportunities, and threats," *Energies*, vol. 11, no. 9, p. 2391, 2018.
12. National Grid plc, "Future energy scenarios," *National Grid Electricity System Operator, UK*, July 2017.
13. H. T. Haider, O. H. See, and W. Elmenreich, "A review of residential demand response of smart grid," *Renewable and Sustainable Energy Reviews*, vol. 59, pp. 166–178, 2016.
14. F. P. Sioshansi, *Variable Generation, Flexible Demand*. Academic Press, Elsevier: USA, 2020.
15. M. Lee *et al.*, "Assessment of demand response and advanced metering," Federal Energy Regulatory Commission, USA, Technical Report, 2013.
16. M. Cheng, S. Sami, and J. Wu, "Virtual energy storage system for smart grids," *Energy Procedia*, vol. 88, pp. 436–442, 2016.
17. M. Cheng, S. S. Sami, and J. Wu, "Benefits of using virtual energy storage system for power system frequency response," *Applied Energy*, vol. 194, pp. 376–385, 2016.
18. P. Ralon, M. Taylor, A. Ilas, H. Diaz-Bone, and K. Kairies, *Electricity Storage and Renewables: Costs and Markets to 2030*, International Renewable Energy Agency: Abu Dhabi, UAE, 2017.
19. E. A. Bhuiyan, M. Z. Hossain, S. Muyeen, S. R. Fahim, S. K. Sarker, and S. K. Das, "Towards next generation virtual power plant: Technology review and frameworks," *Renewable and Sustainable Energy Reviews*, vol. 150, p. 111358, 2021.
20. N. Etherden, V. Vyatkin, and M. H. Bollen, "Virtual power plant for grid services using IEC 61850," *IEEE Transactions on Industrial Informatics*, vol. 12, no. 1, pp. 437–447, 2015.
21. G. Zhang, C. Jiang, and X. Wang, "Comprehensive review on structure and operation of virtual power plant in electrical system," *IET Generation, Transmission & Distribution*, vol. 13, no. 2, pp. 145–156, 2019.
22. M. Cheng, J. Wu, J. Ekanayake, T. Coleman, W. Hung, and N. Jenkins, "Primary frequency response in the Great Britain power system from dynamically controlled refrigerators," 2013.
23. S. S. Sami, M. Cheng and J. Wu, "Modelling and control of multi-type grid-scale energy storage for power system frequency response," *2016 IEEE 8th International Power Electronics and Motion Control Conference, IPEMC-ECCE Asia 2016*, 2016.
24. M. Cheng, J. Wu, S. J. Galsworthy, C. E. Ugalde-Loo, N. Gargov, W. Hung, and N. Jenkins, "Power system frequency response from the control of bitumen tanks," *IEEE Transactions on Power Systems*, vol. 31, no. 3, pp. 1769–1778, 2015, doi: 10.1109/TPWRS.2015.2440336.
25. B. Thomas Alexander, "Technical and commercial integration of distributed and renewable energy sources into existing electricity networks," 2006.
26. M. Cheng, J. Wu, S. Galsworthy, N. Jenkins, and H. William, "Availability of load to provide frequency response in the Great Britain power system," *2014 Power Systems Computation Conference*, pp. 1–7, 2014, doi: 10.1109/PSCC.2014.7038294.
27. BeaconPowerCo, "Beacon Power Technology Brochure." [Online]. Available: http://beaconpower.com/resources/, 2014.

28. S. S. Sami, M. Cheng. J. Wu and N. Jenkins, "A virtual energy storage system for voltage control of distribution networks," *CSEE Journal of Power and Energy Systems*, vol. 4, pp. 146–154, 2018.
29. Statutory Instruments, "The electricity safety, quality and continuity regulations," 2002.
30. Distributed Generation, "Sustainable Electrical Energy Centre. United Kingdom Generic Distribution System (UK GDS)," 2011.

7 Centralized and Decentralized Optimization Approaches for Energy Management within the VPP

Nikos Bogonikolos, Entrit Metai, and Konstantinos Tsiomos
Blockchain2050 BV, CIC Rotterdam, Rotterdam, the Netherlands

CONTENTS

7.1 Introduction 145
    7.1.1 Decentralization – Immutability – Traceability 146
    7.1.2 Smart Contracts 148
    7.1.3 Decentralized Apps (DApps) 149
7.2 Blockchain in the Energy Sector 149
7.3 Significance of Blockchain Technology Incorporation in VPPs 151
7.4 Suggested Methodology 151
7.5 Expected Results 152
7.6 Economic Feasibility of Blockchain Implementation
    in the Energy Sector 153
References 153

7.1 INTRODUCTION

About four decades after the birth of the Internet, and more specifically on January 3, 2009, the first block with 50 Bitcoins was presented by Satoshi Nakamoto. Some people think that this name refers to a group of people and not to an individual. In any case, who he really was remains unknown, and although opinions may differ on his identity, "Satoshi Nakamoto" is considered by many to be a pseudonym.

Bitcoin is the first decentralized digital currency. To be precise, Bitcoin is a cryptocurrency, as its creation uses encryption techniques to verify transactions and

DOI: 10.1201/9781003257202-8

secure the production network. Bitcoin was created to address some of the major drawbacks of conventional trading systems such as vulnerabilities, inadequacies, complexity, and unnecessary additional costs. The previous defects show increasing trends as trading volumes grow. The decentralized nature of Bitcoin ensures that the currency does not have a permanent administrator or a stable authority to control it, as is the case with fiat money that is issued and controlled by central agencies.

Of course, all the previous methods require a technological background to work. This background clearly includes a necessary software but mainly an algorithmic operating framework on which the necessary applications will "run". We could compare it to an operating system in which various software are installed and run. Of course, the algorithmic framework for cryptocurrencies is blockchain. The philosophy behind this technology is based on the following triptych of characteristics.

7.1.1 Decentralization – Immutability – Traceability

Decentralization is the transfer of processes, activities, responsibilities, or powers from a central body to more than one. In the case of blockchain, decentralization can be complete, with all participants being equal and sharing the same set of data at the same time, with each one having their own valid copy. In relation to the physical world, it can be compared to an accounting book (ledger) or some other log file that is distributed to everyone involved and updated simultaneously for everyone. Blockchain as a technology is a type of distributed computing system, which is why it is often referred to as distributed ledger technology (DLT).

Immutability is the blockchain feature that gives it a competitive edge. All data stored on a blockchain network cannot be modified. A transaction, once completed, can only be canceled with a new one. That is, if someone accidentally sends more money to a recipient, then the only way to correct that error is to process a new transaction, from recipient to sender, where the excess amount is refunded. Any attempts to alter the transactions become unsuccessful because blockchain has innate resilient mechanisms for their recovery, which are based on its decentralized structure. This can be illustrated by the following example. Consider that in a transaction, two traders are involved as well as a third with the role of an intermediary. A triplicate receipt is issued for this transaction and each person involved receives a copy. In case one of the three falsifies his/her copy, then it is compared with the other two and its invalidity is ascertained. In such cases, a blockchain network replaces the invalid data with a valid copy. This is where the democracy of blockchain technology comes into play, as the majority prevails. The question then arises: what happens if the majority manages to create a distortion? The answer is that, in this case, the alteration will prevail. But, in order to achieve this, the alteration must be made (almost) at the same time by more than 50% of the participants in the chain. This is called a *51% attack*, and although it is theoretically achievable, in reality, the "attackers" would have to spend vast amounts of processing power in a very short time, which is not realistic.

Through the decentralized blockchain network, data is stored, time-stamped, and can contain additional **traceability** information about the origin of a transaction. Combined with the immutability of the data and the timing they receive when stored

Centralized and Decentralized Optimization Approaches

in a blockchain environment, an unmistakable series of transactions is automatically created that can be traced down to the first block of the chain.

It is clear that the triptych of blockchain technology features **transparency**, which establishes a digital environment of mutual trust resulting from a secure chain containing data of verifiable integrity.

Blockchain is a digital directory in which time-stamped transactions are logged. And although as a transaction, we usually consider the transfer of money (fiat or digital) from one wallet or account to another, it is necessary to remember that in the world of blockchain, the word "transaction" refers to any data exchange that takes place within the network and meets its conditions, whether that exchange refers to financial values or simple information.

Data exchange in a blockchain environment takes place within a peer-to-peer (P2P) network that shares their resources equally, without the need for coordination or mediation by a central server and each participating node has an exact copy of the entire "database" (chain), i.e., all the information shared. Each time a new block of data is added to the chain, the update takes place immediately and simultaneously on all nodes.

When a new chain is launched for the first time, the Genesis Block is born with it. The next trading block will be linked to this. When a participant wants to add a new transaction block to the network, it transmits it to all active nodes. Based on the network verification algorithm, nodes can accept or reject this block. Once validated, it will be saved and added to the chain's most recently saved block and then a copy will be distributed to all participating nodes. This new block will wait for the connection of the next one and this process continues indefinitely. The stored data can also be retrieved to verify an already registered transaction.

For data validation, each blockchain network adopts a consensus mechanism. Each chain uses such an algorithm to validate hashes, with each having its own advantages over the others, such as lower power consumption and shorter validation times.

The decentralized structure of blockchain excludes the existence of intermediaries between traders, thus removing from the transaction any additional costs or restrictions that would arise from a potential intermediaries' involvement. In combination with the immediate processing of transactions, the cost is further reduced as the respective operating costs are reduced.

Blockchain traders can be sure that anything that is circulated, whether it is information about assets or simple data, is logged, accompanied by a time stamp, and then stored in multiple immutable copies, at each node of the chain. Thus, a transparent environment is established where traders have complete control over their assets and are free to exchange any value or data, without restrictions on their size or time. In addition, this environment offers absolute and commonly accepted trust as transactions within it are fully secured, both in terms of fidelity and traceability.

A system enhanced with blockchain technology capabilities (blockchain-enabled) becomes extremely stable, reliable, and secure in terms of the data it hosts and distributes and therefore acquires significant added value. When a blockchain-enabled system is active in the field of trade and the market in general, then it eliminates various frictions that appear in respective conventional systems, such as obstacles

148 Virtual Power Plant Solution for Future Smart Energy Communities

to the exchange of assets, additional charges, financial and time charges (intermediaries, taxes, bureaucratic procedures, etc.), and difficulties in the correct flow of information.

But the capabilities of blockchain technology are not limited to unchanged data storage. There are blockchain environments that allow the unchanging storage of executable code, which launches the usefulness and value of this technology.

7.1.2 SMART CONTRACTS

Smart contracts are executable pieces of code that run in a blockchain environment, each one performing a specific process. Smart contracts work just like conventional contracts, except that their terms and conditions are represented in code. Their general structure is implemented by the If-Then-Else programming flow control commands, which automatically give them flexibility and variety in the range of possibilities they offer. Consequently, a smart contract can be programmed to be a dynamic entity that directly adapts its behavior to existing conditions. Thus, for example, its temporal power may not be constant but may change, always conditionally, and if this has been pre-agreed by the parties involved, before being translated into code.

A smart contract accepts digital data as inputs, which refers to the conditions that will turn it on or off. This data can come from the blockchain network itself but also from external sources such as databases and IoT devices. Each smart contract checks if all its conditions are met. When this happens, then an event is automatically activated by fulfilling its predetermined function. For example, a smart contract could be about collecting data from various sources, and once that collection is complete, then all the data is sent to a predefined email address. These codes serve directly, securely, and transparently, without third-party mediation, and the scenarios that can be served through them are virtually unlimited, ranging from simple transactions (e.g., money transfers or data exchange), to complex processes such as monitoring all stages of a supply chain, with the parallel execution of pre-agreed actions or the integration of game theory methods to maximize the benefit between participants in an activity.

Based on the previous methods, it becomes clear that the flexible nature of smart contracts brings to the table key benefits:

Trust, due to fact that smart contracts' code is stored and runs on a blockchain environment, remaining immutable, and thus all stakeholders can be sure that the code will execute always in a fair manner and for everyone, since the risk of alteration is practically zero.

Security, because all data transactions are encrypted, making it extremely difficult for someone to decrypt them, even if he/she gains access to it. In addition, the link of a data block with the next and previous one renders the whole chain extremely resilient against data tampering.

Accuracy, thanks to the nonexistent human involvement, relevant errors are eliminated, ensuring the seamless execution of contracts anytime, anyplace, for everyone.

Centralized and Decentralized Optimization Approaches 149

Speed, since smart contracts automate procedures, taking them from a human actor and transferring them to a secure digital environment, reducing process times to a minimum.

Cost-effectiveness, by reducing costs through ostracizing unnecessary intermediaries reduces purchases of physical consumables (e.g., paper, printer toners).

Smart contracts are the spearhead of blockchain technology because they are flexible and can be adapted to almost any information need, providing solutions, automating and simplifying processes, while reducing operating costs. Although the use of code with program flow control commands has been a practice for decades in any computer program, the integration of smart contracts in the modern technology landscape is something innovative that further expands the horizons of blockchain technology.

7.1.3 Decentralized Apps (DApps)

Decentralized applications (DApps) are computer applications and are generally open source. Their back-end part is programmed to work in distributed computer system environments, namely in blockchain, while the front-end part is usually designed as a web application. Incoming data can come from a variety of sources, just like smart contracts.

DApps are often referred to as smart contracts. This is partly true. A DApp can be implemented by a single smart contract but large DApps consist of many more, thus covering a wider range of functions. Therefore, we can consider smart contracts to be the building blocks of DApps.

Based on the earlier discussion, it becomes clear that any application which aims to promote trust as an added value by securing data or procedures, smart contracts are the only reliable way to create DApps that will perform the desired functions.

7.2 BLOCKCHAIN IN THE ENERGY SECTOR

The concept of developing a faster, more decentralized web technology systems that are less reliant on human interference has been a driving factor for database innovations. It was an opportunity for blockchain technology that did not go unexploited. Although it emerged as the foundation of cryptocurrencies, its gained momentum exponentially accelerated its spread around the world and blockchain is now globally accepted as a transformative technology. Since its birth, every year that passes, blockchain gains increased acceptance, with several sectors looking eagerly to adopt it as a solution and added value.

Among the various different players, across all industrial sectors, energy-related actors and innovators are well aware that blockchain's potential lies way beyond cryptocurrencies, and they are at the ready to implement this ground-shaking technology to further secure valuable information, automate procedures, and optimize processes through the execution of smart contracts.

The question that comes to mind is how blockchain's full potential can be exploited. The answer is through the effective utilization of smart contracts. As mentioned, these digital contracts execute predefined actions once they are triggered by also a predefined set of conditions. The benefits of smart contracts are nowadays crucial in business collaborations, in which they are mostly utilized to enforce some types of agreements so that all involved participants can be assured and reassured about the process without the involvement of a neutral intermediary.

Currently, we are witnessing the energy sector taking significant evolution steps, transitioning from fossil-fuel-based energy production to a greener production based on renewable sources, with increased storage capabilities in both quality and quantity, as well as optimizing all aspects of energy production and distribution.

Through smart contracts, blockchain technology consists of a solid innovative solution that will deal effectively with the complex processes of energy management because it creates a digital environment of trust, in which data exchanges, payments, agreements, and other transactions between all stakeholders take place securely, faster, smoother, and cheaper.

In the energy sector, it is of utmost importance to increase accuracy, speed, and saving costs without sacrificing security and trust, and for blockchain, all the previous methods are advantageous to the energy sector via the implementation of smart contracts.

Modern energy management systems are quite complex because a wide variety of events that span from production to consumption occur at any given moment. Lack of flexibility to deal with each one may lead to many different types of complications.

By incorporating smart contracts and DApps, an efficient, secure, and seamless distributed energy system can be created to solve the vast majority of issues that may arise. A blockchain-enabled, smart-grid solution for energy distribution management can prove beneficial regardless of the grid scale it serves.

Scalability in blockchain technology is extremely most important, especially in the energy sector where there is a vast quantity of data created and exchanged each second. Monitoring and managing energy values, billing, and communication are just some of the processes that create big volumes of information. Blockchain environments by themselves cannot support such amounts of data within the required timeframes. Luckily, researchers have recommended different approaches to achieve scalability in blockchain, and the most notable of them are the second-layer or off-chain solutions.

These solutions can be regarded as payment channels with huge transaction rates and lightning-fast processing capabilities, while remaining connected to a blockchain. At the end of a transaction in the second layer, the system writes the value back to the main blockchain. For the record, several popular blockchain platforms, such as Bitcoin and Ethereum, have developed off-chain solutions that allow instant transactions between peers. More specifically, Bitcoin has the Lightning Network, while Ethereum has Raiden. Lightning Network is estimated to have a rate of one million transactions per second, while for Raiden, this rate is considered to be infinite. Off-chain solutions have pushed the potential of scaling blockchains to a limitless boundary. What previously was considered a challenge for blockchain is achievable now.

Apart from the high-transaction rates, second-layer solutions bear lower transaction fees compared to the main blockchain as it is independent from the value

Centralized and Decentralized Optimization Approaches 151

transferred, while security and privacy of payments are both maintained. All transactions are executed instantly and the overall capacity of the blockchain network scales linearly with the number of participating.

On the other hand, a token must be deposited before initiating a transaction, which means that one's transactions are limited to the number of tokens owned. Furthermore, one can only transfer to one party per channel; however, multiple channels can be active.

In any case, second-layer solutions make it possible to manage a grid network regardless of its size, without having to worry about transaction latency or throughput. As we speak, blockchain technology has already started transforming the way energy is being handled. Off-chain transactions seem to be the realistic alternative for creating scalable yet sustainable blockchains. Nonetheless, future research projects and collaborations will show whether this solution can be greatly enhanced, particularly in terms of economic value.

7.3 SIGNIFICANCE OF BLOCKCHAIN TECHNOLOGY INCORPORATION IN VPPs

The usage of blockchain in the energy sector has the potential to save costs, improve process efficiency, and increase trade participation. These models also aim to increase prosumers' roles and maximize generation capacity, which can benefit the overall energy ecosystem in the long run. The sharing economy can be enabled and powered by this technology. It is hard to overlook the fact that using blockchain technology to control microgrids and distributed energy resources (DERs) is a viable approach.

Blockchain technology has the potential to speed up grid decarbonization and meet ambitious decarbonization targets in the energy sector by supporting and expanding decentralization and digitalization. As a result, blockchain digitalization will be a tool in attaining the ultimate objective of a speedy energy transition to a world powered entirely by renewable energy sources (RESs).

The application of blockchain in P2P markets is important due to the necessity for decentralized systems to cope with the increasing demand for power and electricity. On their proposed market platform, smart contracts on the blockchain are used to execute such systems, and users can make payments within the blockchain.

7.4 SUGGESTED METHODOLOGY

Smart meters are used to autonomously measure and forecast supply and demand for each consumer, which is subsequently transferred to nodes. To safeguard the privacy of users or consumers, the platform allows this information to be managed locally rather than being sent to the blockchain. The decentralized platform would only assess and disseminate the members' demand and production capacity to other users.

This distributed auction system allows buyers and sellers to conduct dependable, safe, and transparent power transactions. Smart contracts and smart meters are two types of technology used by both buyers and sellers in the system. If the

152 Virtual Power Plant Solution for Future Smart Energy Communities

vendors have surplus generated electricity by the solar cells, they will want to sell it. The procedure is simplified by blockchain technology, and the extra electricity is offered in the grid for purchase by blockchain members. Smart contracts automate the entire process from start to finish. Smart meters in the network detect and report transaction electricity, ensuring that the transaction is completed. All of the data described, which is created by suppliers, bidders, and smart meters, is stored on the blockchain.

A typical blockchain-enabled solution allows consumers to anonymously exchange electricity prices without revealing their identities; as a result, the information exchanged is secure. Certain blockchain technology capabilities, such as anonymous encrypted information transfer and multiple signatures, are used to preserve consumer privacy. Because there is no trustworthy third party in this trade system, prices can be negotiated. Participants remain anonymous at all times, and data security is ensured by robust encryption technologies. A record of the transaction between the two parties is obtained at the end of the process, which is then uploaded to a blockchain and distributed to all participants. This safeguards all participants' privacy, which has been highlighted in earlier studies.

7.5 EXPECTED RESULTS

Blockchain technology is a solid innovative solution that will effectively deal with the complex processes of energy management because it creates a digital environment of trust in which data exchanges, payments, agreements, and other transactions between all stakeholders take place securely, faster, smoother, and cheaper thanks to smart contracts.

It is critical in the energy sector to boost accuracy, speed, and cost savings while maintaining security and trust, and blockchain offers all of these benefits to the energy sector through the deployment of smart contracts.

In the short term, blockchain allows for P2P microgrids, and it will have a long-term impact on electric car integration. Consumer participation in local energy markets could be aided by blockchain technology. As a result, consumers will be exposed to the true cost of energy, which may lead to more efficient energy use or appropriate pricing signals for certain needs. Consumer self-generation investments in photovoltaics (PVs) and small-scale wind energy are examples of local energy markets. Consumers, on the other hand, have not had true access to the energy market until today, which remains a priority for institutionalized energy suppliers due to the high cost of its assets. Furthermore, for the selling of energy surplus back to the grid, various RES investment incentives to export charges are frequently adequate or removed. Utility businesses buy such surpluses at low costs and resell them to consumers at conventional pricing. Saving energy costs for all stakeholders will be attainable if prosumers can directly sell their excess electricity to non-intermediary consumers. Because the earnings and value remain in the microgrid and local communities, prosumers can get more out of their investments. P2P trading can provide alternatives for potential prosumers by acting as social incentives for locally regenerated energy marketplaces.

7.6 ECONOMIC FEASIBILITY OF BLOCKCHAIN IMPLEMENTATION IN THE ENERGY SECTOR

According to economic estimates, establishing a blockchain-supported local energy market from the ground up may be costly, but this cost is likely to drop dramatically as the environment matures. A blockchain-enabled solution, on the other hand, opens up a plethora of trade possibilities and, as a result, might be cost-effective if implemented. The regulatory aspect of implementing such a system should be given special attention: several regulations around the world are not designed for an energy scenario with a high share of distributed production, so it may be more profitable to store energy in batteries rather than selling it in many cases, even if the grid may be better off selling it. Blockchain technology could give some justifications for loosening rules and increasing end-user participation in the market. This is due to smart contracts' neutrality and independence from third parties, which could make it easier to connect end users without the intervention of an energy supplier or another third party.

REFERENCES

Aitzhan, N.Z.; Svetinovic, D. Security and privacy in decentralized energy trading through multi-signatures, blockchain and anonymous messaging streams. IEEE Trans. Dependable Secure Comput. 2018, 15, 840–852.

Brilliantova, V.; Thurner, T.W. Blockchain and the future of energy. Technol. Soc. 2019, 57, 38–45.

Cali, U.; Fifield, A. Towards the decentralized revolution in energy systems using blockchain technology. Int. J. Smart Grid Clean Energy 2019, 8, 245–256.

Cheng, S.; Zeng, B.; Huang, Y.Z. Research on application model of blockchain technology in distributed electricity market. IOP Conf. Ser. Earth Environ. Sci. 2017, 93, 012065.

Department of Energy and Climate Change. Review of the Feed-In Tariffs Scheme; Department of Energy and Climate Change: London, 2015.

Esmat, A.; de Vos, M.; Ghiassi-Farrokhfal, Y.; Palensky, P.; Epema, D. A novel decentralized platform for peer-to-peer energy trading market with blockchain technology. Appl. Energy 2021, 282, 116123.

Hahn, A.; Singh, R.; Liu, C.-C.; Chen, S. Smart contract-based campus demonstration of decentralized transactive energy auctions. In Proceedings of the 2017 IEEE Power & Energy Society Innovative Smart Grid Technologies Conference (ISGT), Washington, DC, USA, 23–26 April 2017; pp. 1–5.

Mengelkamp, E.; Notheisen, B.; Beer, C.; Dauer, D.; Weinhardt, C. A blockchain-based smart grid: Towards sustainable local energy markets. Comput. Sci. Res. Dev. 2018, 33, 207–214.

Mylrea, M.; Gourisetti, S.N.G. Blockchain for smart grid resilience: Exchanging distributed energy at speed, scale and security. In 2017 Resilience Week (RWS); Institute of Electrical and Electronics Engineers (IEEE): Piscataway, NJ, 2017; pp. 18–23.

Park, L.W.; Lee, S.; Chang, H. A sustainable home energy prosumer-chain methodology with energy tags over the blockchain. Sustainability 2018, 10, 658.

Uddin, M.; Romlie, M.F.; Abdullah, M.F.; Halim, S.A.; Kwang, T.C. A review on peak load shaving strategies. Renew. Sustain. Energy Rev. 2018, 82, 3323–3332.

UK Government Chief Scientific Adviser. Distributed Ledger Technology: Beyond Block Chain; UK Government Chief Scientific Adviser: London, 2015.

8 Decision-Making Frameworks for Virtual Power Plant Aggregators in Wholesale Energy and Ancillary Service Markets

S. Bahramara[1] and P. Sheikhahmadi[2]
[1]Department of Electrical Engineering, Sanandaj Branch, Islamic Azad University, Sanandaj, Iran
[2]Department of Electrical and Computer Engineering, University of Kurdistan, Sanandaj, Iran

CONTENTS

8.1 Introduction 155
8.2 Description of the VPPA's Decision-Making Problem in the Markets 158
8.3 Review on Previous Studies.. 159
8.4 Mathematical Formulation 163
8.5 Results.. 166
8.6 Conclusion and Future Works.. 168
References.. 169

8.1 INTRODUCTION

Supplying the world's increasing demand through traditional fossil-fuel-based power plants with low energy efficiency and high emission pollution will be a great threat to global warming. To overcome this problem, the distributed energy resources (DERs) have emerged in the electrical energy systems, which provide two main benefits [1]. First, producing electrical energy from renewable energy-based distributed generations (DGs) decreases the share of electrical energy systems in emission pollution. Second, supplying the electrical demand locally through the DERs can decrease the power loss of the transmission and distribution systems, which consequently decreases the power generation of the power plants. However, managing the high amount of small and medium scales of the DERs, which are mainly installed in the distribution networks, is the main challenge for both the independent system operator (ISO) and the DERs' owners. To solve this challenge, the DERs can be managed by the aggregators as suggested by the International Renewable Energy Agency [2].

DOI: 10.1201/9781003257202-9

Under the management of the aggregators, small-scale DERs would have the opportunity to sell their energy to the wholesale energy markets and they can provide ancillary services for the system operator regarding which they can obtain more profit. Moreover, from its own viewpoint, the ISO only needs to cooperate with a limited number of aggregators in the wholesale markets instead of a large number of the DERs that decrease the complexity of it's decisions. Since integrating a large number of the DERs results in obtaining an appropriate power capacity to participate in the markets, this integration is called *a virtual power plant (VPP) under the management of a VPP aggregator* (VPPA).

The decision-making problem of the VPPAs in the wholesale markets as either the price-taker or the price-maker player is addressed in many studies. The optimal participation of a VPPA in the wholesale real-time (RT) energy market as a price-maker player is addressed in [3] through a bi-level optimization approach. The VPPA's strategies to trade energy with the electric vehicles and responsive loads are modeled through a bi-level optimization model in [4] where the VPPA participates in the wholesale market as a price-taker player. The energy management problem of a VPPA is investigated considering the Internet of Things (IoT) concept in [5] where the VPPA participates in the wholesale day-ahead (DA) and RT energy markets. The strategies of the VPPA to participate in the wholesale energy and reserve markets are mathematically modeled as a robust two-stage stochastic optimization model in [6] to manage the uncertainties of demand and output power of the renewable energy sources (RESs). The bidding strategies of a VPPA in the wholesale energy and regulation markets considering its trading strategies with the DERs' owners are modeled as a tri-level optimization problem in [7]. A risk-based multi-objective optimization model is developed in [8] to model the operation problem of a VPP considering the carbon trading mechanism. In [9], the VPP aims to maximize its overall profit by participating in the bilateral and the DA markets. In fact, an optimal dispatch problem is proposed to increase the weekly VPP's profit associated with the long-term bilateral contract. In [10], a risk-based two-stage stochastic programming is formulated for a VPP with the aim of conducting a daily operation, participating in both the DA and the RT markets. A technical-economic power dispatch is proposed in [11] for a renewable-based VPP participating in the DA energy market. This model is based on a mixed-integer linear programming (MILP) approach and it specifies the hourly operation of a distributed large-scale renewable and on-site plants. A mixed-integer nonlinear programming (MINLP) approach is engaged in [12] to conduct an optimal operation of a VPP besides participating in the DA market while the main purpose is to maximize the operation profit of the VPP. In [13], a techno-economic decision-making framework is addressed for a VPP. In this model, the VPP has an ability to utilize the potential of the DERs to participate simultaneously in both energy and flexible ramping product (FRP) market. In [14], a VPP, as a price-taker player, consisting of photovoltaic (PV) arrays, wind turbines (WTs), and conventional power plant, participates in the bilateral contract and the DA energy markets through an MILP approach. The major objective of this problem is to maximize the VPP's profit. In [15], a VPP, including PV as well as combined heat and power (CHP) systems, is optimally operated through formulating an optimization algorithm. Furthermore, in this study, the VPP participates as a price-taker player in the DA energy market on the one hand and it compensates for the deviations that emerged from the uncertainties of the demand and PVs through participation in the RT market on the other hand. A joint deterministic and interval optimization approach is proposed in [16] to optimize the power dispatch in a VPP participating in the DA market. This model aims at maximizing the VPP's deterministic profit based on the forecast amount of the uncertain parameters on the one hand and maximizing the interval profit regarding the management of the uncertainties on the other hand. In [17], some decision-making frameworks consisting of a robust model, a stochastic programming model, and a robust stochastic programming approach are formulated to provide an energy management model for a VPP participating as a price-taker in the energy markets.

A robust two-stage stochastic programming method is implemented in [18] for the bidding strategy problem of a price-taker VPP in both the DA and the RT energy markets. In the first stage, the VPP specifies the DA decision variables, whereas the RT variables are indicated in the second stage. In [19], the participation of a price-taker VPP containing DGs, electrical energy storage (ESS), flexible demand, and RESs in the DA and RT energy markets is modeled through an adaptive robust stochastic programming method. A novel optimization method is formulated in [20] to model the bidding strategy of a VPP in the energy and reserve markets. The decision-making process in this study is based on strong duality theory and column-and-constraint generation algorithm. In [21], the participation of a VPP consisting of a DG, an electrical energy storage (EES), and a RES in both the DA and RT markets is modeled using a two-stage stochastic programming approach. The main objective of this study is to maximize the expected profit of the VPP. In [22], a Stackelberg game model is proposed to maximize the profit of a VPP participating in the DA and RT markets. In [23], a combination approach, including an adaptive robust optimization and a stochastic programming, is developed with the purpose of modeling the participation of a VPP in the energy and reserve markets. A bi-level stochastic optimization model is formulated in [24] for the participation of a VPP in the DA market. In this model, the upper level problem indicates the maximization of the VPP's profit, whereas the DA market clearing process is formulated in the lower level problem. In [25], a risk-based stochastic programming method is formulated to model the simultaneous bidding strategy of a price-taker VPP in the DA, RT, and reserve markets.

Reviewing the previous studies showed that although the participation problem of the VPPA in the wholesale markets is addressed in many researches, there are still some research gaps in this field, which can be studied by the researchers. For this purpose, in this chapter, different decision-making frameworks are described, which can clarify the VPPAs' participation problem in the wholesale energy and ancillary service markets. Then, a comprehensive review is done on the previous studies from different viewpoints. In the next step, the problem of the VPPA's participation in the markets as a price-taker player is mathematically formulated and the effectiveness of the model is investigated through numerical results. By the end, the conclusion is presented and some future works are described.

The rest of this chapter is organized as follows. The decision-making problem of the VPPA in the markets is described in Section 8.2. A review on the previous studies is done in Section 8.3. The mathematical formulation is presented in Section 8.3.

Numerical results are presented in Section 8.4, and conclusion and future works are given in the last section.

8.2 DESCRIPTION OF THE VPPA'S DECISION-MAKING PROBLEM IN THE MARKETS

The general decision-making framework of the VPPA in the wholesale markets is described in Figure 8.1. It is assumed that the VPP has several resources such as RESs, DGs, and EESs. These resources send their price bids – in order to provide energy and ancillary services – as well as the related constraints to the data control center of the VPPA. The forecasted data of the output power of the RESs and the wholesale market prices is provided by a service provider. It should be noted that the forecasted wholesale market prices are only needed for the price-taker VPPA. The required data of the energy management system of the VPPA is provided by the data control center. In the energy management system, the optimization problem of the VPPA is solved regarding its strategies to participate in the wholesale markets and the output decision variables are sent to the data control center. Then, the bids and constraints of the VPPA are sent to the wholesale markets through the data control center and also this center sends the optimal scheduling signals to the RESs, DGs, and EESs. Different optimization problems of the VPPA in the wholesale markets are described in Figure 8.2. These problems are categorized into four frameworks regarding the VPPA's strategies to participate in the wholesale markets. When the VPPA participates in the markets as a price-taker player, its optimization problem is modeled as frameworks "a" and "b". In these frameworks, the VPPA sends its quantity only bids to the markets to sell energy to the DA and RT markets and provide ancillary services for the market. The objective function of the VPPA is to maximize the profit considering different constraints that are shown in Figure 8.2. The optimization problems of the VPPA are modeled as frameworks "c" and "d" in cases where the aggregator participates in the markets as a price-maker player. In these frameworks, the VPPA sends the bids price to sell energy to the energy markets and provide ancillary services for the system;

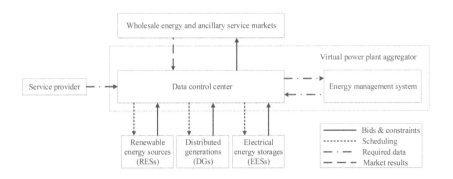

FIGURE 8.1 Decision-making framework of the VPPA in the wholesale markets.

Decision-Making Frameworks for Virtual Power Plant Aggregators

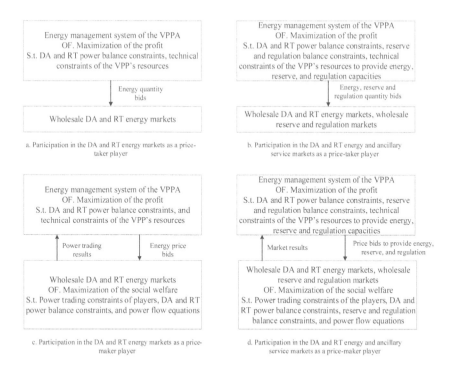

FIGURE 8.2 Different optimization problems of the VPPA.

therefore, the wholesale markets are cleared and the market results are announced to the VPPA.

8.3 REVIEW ON PREVIOUS STUDIES

A comprehensive review on some previous studies in the field of the participation of the VPPA in the wholesale markets is given in Tables 8.1 and 8.2. The type of DERs used in the VPPs on the one hand and the strategies of the VPPA to participate in different markets on the other hand are reviewed in Table 8.2. In these tables, the symbol * shows the existence of the item (such as type of DERs or type of markets) in the references. As shown in this table, all the DERs consisting of RESs, DGs, EESs, and demand response programs (DRPs) are modeled in the VPPs. However, the CHP units are modeled in a few studies to meet both electrical and thermal demands. Also, the results showed that the VPPs mainly participate in the markets as the price-taker players and their strategies as the price-maker players are modeled in few ones. From the viewpoint of participating in different markets, the VPPs are mainly modeled in the DA and RT energy markets and participating in the ancillary service markets is considered in limited studies. This occurs when the flexible energy resources such as the EESs and DGs can provide ability for the VPPA to provide ancillary services for the system.

TABLE 8.1

Strategies of the VPPs Equipped with DERs to Participate in Different Markets

Ref.	Type of DERs						Type of Participation in Markets		Type of Markets			
									Energy		Ancillary Service	
	PV	WT	DG	EES	DRP	CHP	Price-taker	Price-maker	DA	RT	Reserve	Regulation
[3]	*	*	*	*	-	-	-	*	-	*	-	-
[4]	-	*	-	*	*	-	*	-	*	-	-	*
[5]	*	-	-	*	*	-	*	-	*	*	-	-
[6]	*	*	*	-	-	*	*	-	*	-	*	-
[7]	*	*	*	*	*	-	-	*	*	-	-	*
[8]	*	*	*	*	*	*	*	-	*	-	-	-
[9]	*	*	*	*	-	-	-	*	*	*	-	-
[10]	*	*	-	-	-	-	-	*	*	-	-	-
[11]	*	*	-	*	-	-	*	-	*	*	-	-
[12]	*	*	*	*	-	-	*	-	*	-	-	*
[13]	*	-	*	*	*	-	*	-	*	-	*	-
[14]	*	*	-	*	-	-	*	-	*	*	-	-
[15]	*	-	-	*	-	*	*	-	*	-	-	-
[16]	-	*	-	*	*	*	*	-	*	*	-	-
[17]	-	*	*	*	-	-	*	-	*	*	-	-
[18]	-	*	*	*	*	-	*	-	*	*	-	-
[19]	*	*	*	*	*	-	*	-	*	*	-	-
[20]	-	-	-	*	-	-	*	-	*	*	*	-
[21]	*	*	*	*	*	-	*	-	*	*	-	-
[22]	-	-	-	*	*	-	*	-	*	-	-	-
[23]	*	*	*	*	*	-	*	-	*	-	*	-
[24]	-	*	-	*	*	-	-	*	*	-	-	-
[25]	*	*	*	*	-	-	*	-	*	*	*	-

Decision-Making Frameworks for Virtual Power Plant Aggregators   161

TABLE 8.2

Different Models Proposed for the VPPs

Ref.	Uncertain Parameters			Method of Modeling Uncertainties	Type of Model	Solution Method
	Output Power of RESs	Demand	Price			
[3]	*	-	-	Two-stage stochastic programming	MILP	GAMS
[4]	*	*	*	Stochastic programming	MILP	GAMS
[5]	-	*	-	Historical data	MILP	Nod-Red programming tool
[6]	*	*	*	Two-stage stochastic programming	MILP	Heuristic algorithms
[7]	*	*	-	Deep learning based (PQC forecasting)	MILP	Mathematical package
[8]	*	*	*	Historical data	NLP	Heuristic algorithms
[9]	*	*	*	Historical data/ stochastic programming	MILP/ branch-and-bound	GAMS
[10]	*	-	*	Two-stage stochastic programming	MILP/ branch-and-bound	GAMS
[11]	-	-	-		MILP/ branch-and-bound	MATLAB
[12]	-	-	*	Historical data	MINLP/ branch-and-bound	LINGO
[13]	-	-	*	Historical data	MILP	Heuristic algorithms
[14]	*	-	-	Historical data	Branch-and-bound	Heuristic algorithms
[15]	*	*	*	Historical data	MILP	Heuristic algorithms
[16]	*	-	-	Historical data (interval)	MIP	Commercial solvers (CPLEX)
[17]	*	-	*	Stochastic programming/ historical data (prediction interval)	MILP/ branch-and-bound	GAMS
[18]	*	-	*	Two-stage stochastic programming	LP/simplex	Mathematical method

(Continued)

TABLE 8.2 *(Continued)*
Different Models Proposed for the VPPs

| | Uncertain Parameters | | | | | |
Ref.	Output Power of RESs	Demand	Price	Method of Modeling Uncertainties	Type of Model	Solution Method
[19]	*	-	*	Confidence bounds and scenarios/ stochastic programming	MILP/ branch-and-bound	GAMS
[20]	*	-	*	Stochastic programming	MILP/column generation	GAMS
[21]	*	-	*	Historical data	MILP/ branch-and-bound	GAMS
[22]	*	*	-	Monte Carlo	MILP/game theory	YALMIP/ MATLAB
[23]	*	-	*	Confidence bounds and intervals/ stochastic programming	Column-and-constraint generation	Heuristic algorithms
[24]	*	*	-	Stochastic programming	MILP	GAMS
[25]	*	*	*	Two-stage stochastic programming	MILP	GAMS

Different modeling approaches of the VPP's decision-making problem in the markets are reviewed in Table 8.3. As shown in this table, the uncertainties related to output power of the RESs, demand, and energy prices are modeled through different models in the previous studies. Using the historical data and modeling the uncertainties through the stochastic optimization approach are main approaches used to

TABLE 8.3
Bids and Technical Constraints of the DERs (Cost Unit Is $/MWh and Power Unit Is MW)

# DG	$\overline{P}_k^{DG}$	$\underline{P}_k^{DG}$	RU_k	RD_k	P_{DG}^{ini}	C_k^{DG}	$C_k^{DG\_Re}$
1, 2	0.5	0	0.30	0.30	0.2	10	3.0
3, 4	1.0	0	0.70	0.70	0.3	13	3.9
# ES	$\overline{P}_e^{ch}/\overline{P}_e^{dch}$	$\underline{E}_e$	$\overline{E}_e$	η_{ch}, η_{dch}	E_e^{ini}	C_e^{ESch}/C_e^{ESdch}	$C_e^{ES\_Re}$
1, 2	0.5	1	2.5	0.95	1	2.5	0.75
3, 4	0.5	1	2.5	0.90	1	3.0	1.00

Decision-Making Frameworks for Virtual Power Plant Aggregators   163

model the uncertainties in these studies. Also, the VPPs' trading strategies in the markets are mainly modeled as the MILP and MINLP models and they are solved through the GAMS software in many studies.

8.4 MATHEMATICAL FORMULATION

In this section, a basic formulation is presented for the participation of the VPPA in the wholesale DA and RT energy and the reserve markets as a price-taker player.[1] The objective function of the VPPA is formulated in Equation (8.1).

$$TP^{\text{VPP}} = R^{\text{DA}} + R^{\text{RT}} + R^{\text{Res.}} - C^{\text{DA}} - C^{\text{RT}} - C^{\text{Res.}} \tag{8.1}$$

where R^{DA}, R^{RT}, and $R^{\text{Res.}}$ show the revenue of the VPPA from participating in the DA and RT energy and the reserve markets, respectively, which are modeled in Equation (8.2). Also, C^{DA}, C^{RT}, and $C^{\text{Res.}}$ are used to show the cost of providing energy and reserve by the DERs for the VPPA, which are modeled in Equations (8.3) and (8.4).

$$R^{\text{DA}} = \sum_{t=1}^{T} \lambda_t^{\text{DA}} p_t^{\text{DA}}, \qquad R^{\text{RT}} = \sum_{t=1}^{T} \lambda_t^{\text{RT}} p_t^{\text{RT}},$$

$$R^{\text{Res.}} = \sum_{t=1}^{T} \left(\lambda_t^{\text{Res.}} p_t^{\text{Res.}} + \lambda_t^{\text{RT}} p_t^{\text{Res.}} \kappa_t^{\text{Res.}} \right) \tag{8.2}$$

In Equation (8.2), λ^{DA}, λ^{RT}, and $\lambda^{\text{Res.}}$ are the price of the DA, RT, and reserve wholesale markets, respectively. In this equation, p_t^{DA} and p_t^{RT} show the sold power of the VPPA to the DA and RT energy markets and $p_t^{\text{Res.}}$ is used to model the reserve provided by the VPPA for the market. Also, $\kappa_t^{\text{Res.}}$ is the probability of calling reserve in the market in each time step. As shown in Equation (8.2), there are two terms in the revenue of the VPPA from participating in the reserve market: (1) Revenue from providing the reserve capacity in the DA scheduling and (2) revenue from providing energy in the RT operation when the reserve capacity is called by the ISO.

$$C^{\text{DA}} = \sum_{t=1}^{T} \left[\sum_{i=1}^{I} \lambda_{i,t}^{\text{DG}} p_{i,t}^{\text{DG\_DA}} + \sum_{n=1}^{N} \left(\lambda_{n,t}^{\text{dis.}} p_{n,t}^{\text{dis\_DA}} - \lambda_{n,t}^{\text{ch.}} p_{n,t}^{\text{ch\_DA}} \right) \right] \tag{8.3}$$

$$C^{\text{RT}} = \sum_{t=1}^{T} \left[\sum_{i=1}^{I} \lambda_{i,t}^{\text{DG}} p_{i,t}^{\text{DG\_RT}} + \sum_{n=1}^{N} \left(\lambda_{n,t}^{\text{dis.}} p_{n,t}^{\text{dis\_RT}} - \lambda_{n,t}^{\text{ch.}} p_{n,t}^{\text{ch\_RT}} \right) \right] \tag{8.4}$$

In these equations, $\lambda_{i,t}^{\text{DG}}$, $\lambda_{n,t}^{\text{dis.}}$, and $\lambda_{n,t}^{\text{ch.}}$ are the bid of the DGs and the EESs to trade energy with the VPPA. $p_{i,t}^{\text{DG\_DA}}$ and $\lambda_{i,t}^{\text{DG\_RT}}$ are the power generation of the DGs in the DA and the RT operation. $p_{i,t}^{\text{dis\_DA}}$ and $\lambda_{i,t}^{\text{dis\_RT}}$ are the discharging power of the EESs in the DA and RT operation, respectively. Also, the charging power of the EESs in the DA and RT are shown as $p_{i,t}^{\text{ch\_DA}}$ and $\lambda_{i,t}^{\text{ch\_RT}}$, respectively. Since the EESs' owners purchase energy from the VPPA to charge their EESs, this term is modeled as a revenue term in Equations (8.3) and (8.4) from the viewpoint of the VPPA.

$$
\begin{aligned}
C^{\text{Res.}} = \sum_{t=1}^{T} \Bigg[& \sum_{i=1}^{I} \lambda_{i,t}^{\text{DG\_Res.}} \, p_{i,t}^{\text{DG\_Res.}} + \lambda_{i,t}^{\text{DG}} \, p_{i,t}^{\text{DG\_Res.}} \, \kappa_t^{\text{Res.}} \\
& + \sum_{n=1}^{N} \lambda_{n,t}^{\text{b\_Res.}} \, p_{n,t}^{\text{b\_Res.}} + \lambda_{n,t}^{\text{dis.}} \, p_{n,t}^{\text{b\_Res.}} \, \kappa_t^{\text{Res.}} \Bigg]
\end{aligned}
\tag{8.5}
$$

where $\lambda_{i,t}^{\text{DG\_Res.}}$ and $\lambda_{n,t}^{\text{b\_Res.}}$ are the bids of the DGs and the EESs to provide reserve for the VPPA, respectively. Also, $p_{i,t}^{\text{DG\_Res.}}$ and $p_{n,t}^{\text{b\_Res.}}$ are the amount of the reserve capacity provided by the DGs and the EESs for the VPPA. The VPPA pays two cost terms to the owners of DGs and EESs for providing the reserve capacity in the DA scheduling (the first and the third terms of Equation (8.5)) and providing energy in the RT dispatching when the reserve capacity is called by the system operator (the second and fourth terms of Equation (8.5)).

This objective function is maximized considering the following constraints:

$$
\sum_{i=1}^{I} p_{i,t}^{\text{DG\_DA}} + \sum_{n=1}^{N} p_{n,t}^{\text{dis\_DA}} = \sum_{n=1}^{N} p_{n,t}^{\text{ch\_DA}} + p_t^{\text{DA}}
\tag{8.6}
$$

$$
\begin{aligned}
& \sum_{i=1}^{I} \left(p_{i,t}^{\text{DG\_DA}} + p_{i,t}^{\text{DG\_RT}} + \kappa_t^{\text{Res.}} \, p_{i,t}^{\text{DG\_Res.}} \right) \\
& + \sum_{n=1}^{N} \left(p_{n,t}^{\text{dis\_DA}} + p_{n,t}^{\text{dis\_RT}} + \kappa_t^{\text{Res.}} \, p_{n,t}^{\text{b\_Res.}} \right) \\
& = \sum_{n=1}^{N} \left(p_{n,t}^{\text{ch\_DA}} + p_{n,t}^{\text{ch\_RT}} \right) + \left(p_t^{\text{DA}} + p_t^{\text{RT}} + \kappa_t^{\text{Res.}} \, p_t^{\text{Res.}} \right)
\end{aligned}
\tag{8.7}
$$

Equations (8.6) and (8.7) are used to model the DA and RT energy balance constraint of the VPPA, respectively, where in both equations, the sum of the power generation of the DGs and discharging power of EESs is equal to the sum of the charging power of the EESs and the sold power to the energy markets.

$$
p_t^{\text{Res.}} = \sum_{i=1}^{I} p_{i,t}^{\text{DG\_Res.}} + \sum_{n=1}^{N} p_{n,t}^{\text{b\_Res.}}
\tag{8.8}
$$

The reserve capacity balance constraint of the VPPA is modeled in Equation (8.8) where the sum of the reserve provided by the DGs and EESs for the VPPA is equal to the reserve capacity provided by the VPPA for the market.

$$
p_{i,t}^{\text{DG\_DA}} + p_{i,t}^{\text{DG\_Res.}} \le \overline{P}_{i,t}^{\text{DG}}, \qquad p_{i,t}^{\text{DG\_DA}} \ge 0, \qquad p_{i,t}^{\text{DG\_Res.}} \ge 0
\tag{8.9}
$$

Decision-Making Frameworks for Virtual Power Plant Aggregators 165

$$p_{i,t}^{\text{DG\_DA}} + p_{i,t}^{\text{DG\_Res.}} - p_{i,t-1}^{\text{DG\_DA}} \leq \text{RU}_i,$$

$$p_{i,t-1}^{\text{DG\_Res.}} + p_{i,t-1}^{\text{DG\_DA}} - p_{i,t}^{\text{DG\_DA}} \leq \text{RD}_i \tag{8.10}$$

$$p_{i,t}^{\text{DG\_DA}} + p_{i,t}^{\text{DG\_RT}} + p_{i,t}^{\text{DG\_Res.}} \kappa_t^{\text{Res.}} \leq \overline{P}_{i,t}^{\text{DG}}, \qquad p_{i,t}^{\text{DG\_RT}} \geq 0 \tag{8.11}$$

$$p_{i,t}^{\text{DG\_DA}} + p_{i,t}^{\text{DG\_RT}} + p_{i,t}^{\text{DG\_Res.}} \kappa_t^{\text{Res.}} - p_{i,t-1}^{\text{DG\_DA}} - p_{i,t-1}^{\text{DG\_RT}} \leq \text{RU}_i \tag{8.12}$$

$$p_{i,t-1}^{\text{DG\_DA}} + p_{i,t-1}^{\text{DG\_RT}} + p_{i,t-1}^{\text{DG\_Res.}} \kappa_{t-1}^{\text{Res.}} - p_{i,t}^{\text{DG\_DA}} - p_{i,t}^{\text{DG\_RT}} \leq \text{RD}_i \tag{8.13}$$

Equations (8.9)–(8.13) are used to model the technical constraints of the DGs to provide energy and reserve for the VPPA. Equation (8.9) shows that the sum of the DA energy generation of the DGs and their reserve capacity provided for the VPPA is lower than or equal to the maximum capacity of DGs ($\overline{P}_{i,t}^{\text{DG}}$). The ramp-up (RU) and ramp-down (RD) limitations of DGs in its DA scheduling are modeled in Equation (8.10). The technical constraints of the DGs in the RT operation are modeled in Equations (8.11)–(8.13) considering the probability of calling reserve capacity in the RT operation.

$$0 \leq p_{n,t}^{\text{ch\_DA}} \leq \overline{P}_{n,t}^{\text{b}} U_{n,t}^{\text{b\_DA}},$$

$$0 \leq p_{n,t}^{\text{dis\_DA}} \leq \overline{P}_{n,t}^{\text{b}} (1 - U_{n,t}^{\text{b\_DA}}) \tag{8.14}$$

$$p_{n,t}^{\text{dis\_DA}} + p_{n,t}^{\text{b\_Res.}} - p_{n,t}^{\text{ch\_DA}} \leq \overline{P}_{n,t}^{\text{b}} \tag{8.15}$$

$$e_{n,t}^{\text{b\_DA}} = e_{n,t-1}^{\text{b\_DA}} + p_{n,t}^{\text{ch\_DA}} \eta_n^{\text{ch.}} - \frac{p_{n,t}^{\text{dis\_DA}}}{\eta_n^{\text{dis.}}} \tag{8.16}$$

$$\underline{E}_n^{\text{b}} \leq e_{n,t}^{\text{b\_DA}} \leq \overline{E}_n^{\text{b}} \tag{8.17}$$

$$p_{n,t}^{\text{b\_Res.}} \leq e_{n,t}^{\text{b\_DA}} - \underline{E}_n^{\text{b}} \tag{8.18}$$

$$0 \leq p_{n,t}^{\text{ch\_RT}} \leq \overline{P}_{n,t}^{\text{b}} U_{n,t}^{\text{b\_RT}},$$

$$0 \leq p_{n,t}^{\text{dis\_RT}} \leq \overline{P}_{n,t}^{\text{b}} (1 - U_{n,t}^{\text{b\_RT}}) \tag{8.19}$$

$$p_{n,t}^{\text{ch\_DA}} + p_{n,t}^{\text{ch\_RT}} - p_{n,t}^{\text{b\_Res.}} \kappa_t^{\text{Res.}} \leq \overline{P}_{n,t}^{\text{b}} \tag{8.20}$$

$$p_{n,t}^{\text{dis\_DA}} + p_{n,t}^{\text{dis\_RT}} + p_{n,t}^{\text{b\_Res.}} \kappa_t^{\text{Res.}} \leq \overline{P}_{n,t}^{\text{b}} \tag{8.21}$$

$$e_{n,t}^{b\_RT} = e_{n,t-1}^{b\_RT} + \left(p_{n,t}^{ch\_DA} + p_{n,t}^{ch\_RT}\right)\eta_n^{ch.}$$
$$- \frac{\left(p_{n,t}^{dis\_DA} + p_{n,t}^{dis\_RT} + p_{n,t}^{b\_Res.} \kappa_t^{Res.}\right)}{\eta_n^{dis.}} \quad (8.22)$$

$$\underline{E}_n^b \leq e_{n,t}^{b\_RT} \leq \overline{E}_n^b \quad (8.23)$$

Equations (8.14)–(8.23) are used to model the technical constraints of the EESs in the DA and RT operation. The maximum power charging/discharging limitations of the EESs in the DA and RT operation are modeled in Equations (8.14) and (8.19), respectively. In these equations, the binary variables $U_{n,t}^{b\_DA}$ and $U_{n,t}^{b\_RT}$ avoid from charging and discharging the EESs simultaneously. Equations (8.15), (8.20), and (8.21) are used to model the power limitations of the EESs to provide the reserve capacity for the VPPA and its deployment in the RT operation. Equations (8.16) and (8.22) are used to model the energy balance constraint of the EESs in the DA and RT operation. Also, the minimum and maximum limitations of the stored energy in the EESs in the DA and RT are modeled in Equations (8.17) and (8.23), respectively. The energy limitation of the EESs to provide the reserve capacity in the DA scheduling is modeled in Equation (8.18).

8.5 RESULTS

The behavior of the VPP in the DA, RT, and reserve market is investigated in this section. The bids of the DERs and their technical constraints are indicated in Table 8.3 [26, 27]. The DA and RT energy market prices as well as the reserve market price are presented in Figures 8.3 and 8.4, respectively [28]. The probability of the reserve capacity deployment is set at 0.2.

The revenue of the VPP from participating in the DA, RT, and reserve markets are 505.38 $, 1390.66 $, and 193.76 $, respectively. Moreover, the total profit of the revenue from participating in these markets is 1317.59 $. In the case that the VPP only participates in the DA and RT energy markets, its revenue is 1222.37 $. Therefore, participating in the reserve market increases the revenue of the VPP.

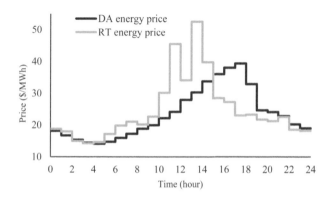

FIGURE 8.3 The DA and RT energy markets prices.

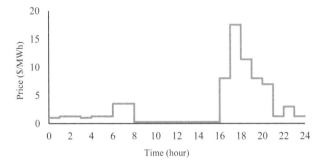

FIGURE 8.4 The reserve market price.

The DA and RT decisions of the VPP in the markets and its decisions on the optimal scheduling of the resources are shown in Figures 8.5–8.7. The VPP decides to sell energy to the DA and RT markets regarding the energy market prices. In fact, the VPP sells energy to the RT energy market in hours when the RT energy market price is more than the DA one. For instance, it sells energy to the RT market in hours 1–2 and 5–15 owing to higher market prices in comparison with the DA market. In contrast, regarding a similar reason, the VPP sells energy to the DA market in hours 3–4, 16–19, and 21–24. As shown in Figure 8.6, the VPP provides the reserve capacity for the market in hours 3, 5–8, 17–21, and 23 regarding the high reserve market price in these hours. It is notable that the VPP imposes a different strategy in hour 20 to simultaneously participate in energy and reserve markets. For this purpose, the VPP charges the ESSs using the DGs regarding which the VPP has an ability to provide the maximum amount of reserve capacity for the reserve market, i.e. 5 MW (by charging the EES, its ability to provide more reserve capacity increases as shown in Equation (8.20)). On the other hand, there is some part of the DGs' capacity based on the reserve deployment probability regarding which the VPP can also sell the energy to the RT market in this hour.

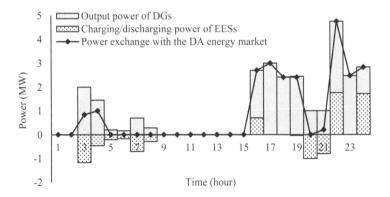

FIGURE 8.5 Power balance in the DA operation.

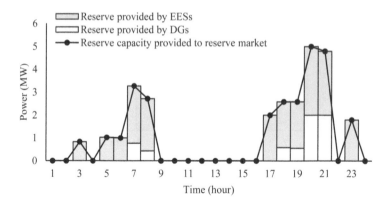

FIGURE 8.6 The VPP's decisions to provide the reserve capacity.

8.6 CONCLUSION AND FUTURE WORKS

Integration of the DERs into VPPs has benefits for both the DERs and the system operators. Since the aim of the VPPAs is to obtain profit from participating in the wholesale energy and ancillary service markets, in this chapter, appropriate decision-making frameworks are proposed for the VPPA to participate in the markets. In these frameworks, two strategies of the VPPAs are described in their participation in the markets; price-taker and price-maker strategies. Then, some of the previous studies that model the VPPA's decisions in the wholesale markets are reviewed. A general formulation is developed to model the VPPA's optimization problem to participate in the wholesale DA and RT energy and reserve markets as a price-taker player. In the end, the behavior of the VPP in these markets is analyzed through numerical results.

As shown in this chapter, there are still some research gaps in this field, which can be studied by the researchers as the future works and some of them are proposed as follows:

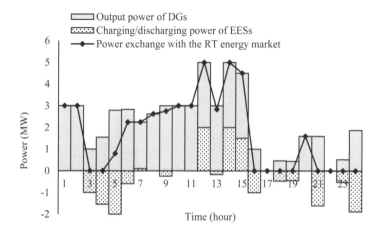

FIGURE 8.7 The VPP's decisions in the RT operation.

Decision-Making Frameworks for Virtual Power Plant Aggregators

- Modeling the decision-making problem of the VPPAs in the wholesale energy and ancillary service markets as the price-maker players
- Proposing new mathematical models for the decision-making problem of the VPPA in the ancillary service markets
- Coupling the DA scheduling and the RT dispatching of the DERs in the VPP considering the different decision-making strategies of the VPPA in the energy and ancillary service markets
- Using the data-driven approaches to model the uncertainties of the RESs, demand, and price in the VPPA's optimization problem

NOTE

1. For further details about the formulation of the VPP's problem in the regulation market or the bidding strategies of the VPPA in the markets, please see the related references reviewed in Section 8.1.

REFERENCES

1. S. Bahramara, A. Mazza, G. Chicco, M. Shafie-khah, J.P.S. Catalão, Comprehensive review on the decision-making frameworks referring to the distribution network operation problem in the presence of distributed energy resources and microgrids, International Journal of Electrical Power & Energy Systems, 115 (2020) 105466.
2. International Renewable Energy Agency, Innovation landscape for a renewable-powered future, International Renewable Energy Agency, Abu Dhabi (2019a).
3. P. Sheikhahmadi, S. Bahramara, The participation of a renewable energy-based aggregator in real-time market: A bi-level approach, Journal of Cleaner Production, 276 (2020) 123149.
4. H. Rashidizadeh-Kermani, M. Vahedipour-Dahraie, M. Shafie-Khah, P. Siano, A stochastic short-term scheduling of virtual power plants with electric vehicles under competitive markets, International Journal of Electrical Power & Energy Systems, 124 (2021) 106343.
5. P. Pal, A. Parvathy, K. Devabalaji, S.J. Antony, S. Ocheme, T.S. Babu, H.H. Alhelou, T. Yuvaraj, IoT-based real time energy management of virtual power plant using PLC for transactive energy framework, IEEE Access, 9 (2021) 97643–97660.
6. F. Fang, S. Yu, X. Xin, Data-driven-based stochastic robust optimization for a virtual power plant with multiple uncertainties, IEEE Transactions on Power Systems, 12 (1) (2022) 456–466.
7. Z. Yi, Y. Xu, H. Wang, L. Sang, Coordinated operation strategy for a virtual power plant with multiple DER aggregators, IEEE Transactions on Sustainable Energy, 12 (2021) 2445–2458.
8. C. Tan, J. Wang, S. Geng, L. Pu, Z. Tan, Three-level market optimization model of virtual power plant with carbon capture equipment considering copula–CVaR theory, Energy, 237 (2021) 121620.
9. H. Pandžić, I. Kuzle, T. Capuder, Virtual power plant mid-term dispatch optimization, Applied Energy, 101 (2013) 134–141.
10. M.A. Tajeddini, A. Rahimi-Kian, A. Soroudi, Risk averse optimal operation of a virtual power plant using two stage stochastic programming, Energy, 73 (2014) 958–967.
11. N. Naval, R. Sánchez, J.M. Yusta, A virtual power plant optimal dispatch model with large and small-scale distributed renewable generation, Renewable Energy, 151 (2020) 57–69.

12. N. Naval, J.M. Yusta, Water-energy management for demand charges and energy cost optimization of a pumping stations system under a renewable virtual power plant model, Energies, 13 (2020) 2900.
13. H. Wang, S. Riaz, P. Mancarella, Integrated techno-economic modeling, flexibility analysis, and business case assessment of an urban virtual power plant with multimarket co-optimization, Applied Energy, 259 (2020) 114142.
14. M. Zdrilić, H. Pandžić, I. Kuzle, The mixed-integer linear optimization model of virtual power plant operation, in: 2011 8th International Conference on the European Energy Market (EEM), IEEE, 2011, pp. 467–471.
15. J. Zapata, J. Vandewalle, W. D'haeseleer, A comparative study of imbalance reduction strategies for virtual power plant operation, Applied Thermal Engineering, 71 (2014) 847–857.
16. Y. Liu, M. Li, H. Lian, X. Tang, C. Liu, C. Jiang, Optimal dispatch of virtual power plant using interval and deterministic combined optimization, International Journal of Electrical Power & Energy Systems, 102 (2018) 235–244.
17. M. Rahimiyan, L. Baringo, Real-time energy management of a smart virtual power plant, IET Generation, Transmission & Distribution, 13 (2019) 2015–2023.
18. M. Rahimiyan, L. Baringo, Strategic bidding for a virtual power plant in the day-ahead and real-time markets: A price-taker robust optimization approach, IEEE Transactions on Power Systems, 31 (2015) 2676–2687.
19. A. Baringo, L. Baringo, A stochastic adaptive robust optimization approach for the offering strategy of a virtual power plant, IEEE Transactions on Power Systems, 32 (2016) 3492–3504.
20. Y. Zhou, Z. Wei, G. Sun, K.W. Cheung, H. Zang, S. Chen, Four-level robust model for a virtual power plant in energy and reserve markets, IET Generation, Transmission & Distribution, 13 (2019) 2036–2043.
21. H. Pandžić, J.M. Morales, A.J. Conejo, I. Kuzle, Offering model for a virtual power plant based on stochastic programming, Applied Energy, 105 (2013) 282–292.
22. H. Wu, X. Liu, B. Ye, B. Xu, Optimal dispatch and bidding strategy of a virtual power plant based on a Stackelberg game, IET Generation, Transmission & Distribution, 14 (2019) 552–563.
23. A. Baringo, L. Baringo, J.M. Arroyo, Day-ahead self-scheduling of a virtual power plant in energy and reserve electricity markets under uncertainty, IEEE Transactions on Power Systems, 34 (2018) 1881–1894.
24. E.G. Kardakos, C.K. Simoglou, A.G. Bakirtzis, Optimal offering strategy of a virtual power plant: A stochastic bi-level approach, IEEE Transactions on Smart Grid, 7 (2015) 794–806.
25. M. Vahedipour-Dahraie, H. Rashidizadeh-Kermani, M. Shafie-Khah, J.P. Catalão, Risk-averse optimal energy and reserve scheduling for virtual power plants incorporating demand response programs, IEEE Transactions on Smart Grid, 12 (2020) 1405–1415.
26. W. Alharbi, K. Raahemifar, Probabilistic coordination of microgrid energy resources operation considering uncertainties, Electric Power Systems Research, 128 (2015) 1–10.
27. S. Bahramara, M.P. Moghaddam, M.R. Haghifam, Modelling hierarchical decision making framework for operation of active distribution grids, IET Generation, Transmission & Distribution, 9 (2015) 2555–2564.
28. J. Wang, H. Zhong, W. Tang, R. Rajagopal, Q. Xia, C. Kang, Y. Wang, Optimal bidding strategy for microgrids in joint energy and ancillary service markets considering flexible ramping products, Applied Energy, 205 (2017) 294–303.

9 Decentralized Energy Management System within VPP

K. Shanti Swarup and P. M. Naina
Department of Electrical Engineering, Indian
Institute of Technology Madras, Chennai, India

CONTENTS

9.1 Introduction 172
9.2 Virtual Power Plant Model 174
    9.2.1 Virtual Power Plant Concept 174
9.3 Energy Management System For VPP 175
9.4 Decentralized Energy Management System 175
9.5 Problem Formulation 177
9.6 Consensus-Based Distributed Algorithm 178
    9.6.1 The First-Order Discrete-Time Consensus Protocol 178
9.7 Case Study 180
9.8 Simulation Results and Discussions 181
    9.8.1 Evaluating Small VPP, VPP1 181
    9.8.2 Evaluating Large VPP, VPP2 183
        9.8.2.1 Case Study I – No Power Exchange Between VPP
             and the Grid 183
        9.8.2.2 Case Study II – No Power Exchange Between VPP
             and the Grid Considering Controllable Loads 185
        9.8.2.3 Case Study III – Power Export Mode of VPP 185
        9.8.2.4 Case Study IV – Power Import Mode of VPP 185
        9.8.2.5 Case Study V – VPP Operation with Communication
             Delay and Noise 186
9.9 Conclusion 188
References 188

NOMENCLATURE

N no. of nodes
ρ_E price at which VPP buys electricity from the market
ρ_s price at which VPP sells electricity to its customers
P_s power injected into VPP

DOI: 10.1201/9781003257202-10

P_{dn}	load demand at the nth node
P_{dgn}	power output of DG at the nth node
P_{dgn}^{max}	upper limit of the DG generation at the nth node
P_{dgn}^{min}	lower limit of the DG generation at the nth node
P_{cn}	power curtailed by the controllable load at the node n
P_{cn}^{max}	upper limit of the controllable load at the nth node
τ	communication delay
σ	noise
$\bar{d}$	maximum delay applied
D	doubly stochastic matrix
L	Laplacian matrix of the graph
$c(k)$	consensus gain function
k	iteration index
λ	incremental cost
ΔP	power imbalance
a, b	cost coefficients of the quadratic cost function

9.1 INTRODUCTION

The high penetration of renewable sources into the existing grid increases the complexities of the grid operation. The power generation from distributed generators will soon hold a major share of power consumption. Nowadays, the electricity markets have become significant since markets enable competition among the participants, which lowers the energy cost. But the renewable sources cannot participate in the electricity markets due to their smaller capacity and uncertain nature. There are three possible paradigms for designing energy markets to manage these distributed resources, i.e., peer-to-peer, prosumer-to-grid, and community-based schemes [1]. Virtual power plant (VPP) is a particular case of the community-based approach, which integrates prosumers from different geographical locations. The main objective of VPP is to encourage distributed energy resources (DERs')/prosumers' participation by improving their profits from the energy market. The VPP is considered as a control center that communicates with the Independent System Operator (ISO) and also participates in electricity markets. Nowadays, VPPs are participating in real-time markets as well as day-ahead markets.

The VPP is similar to a conventional power plant [2]. VPP aggregates the capacity of its distributed generating units [3]. One of the main advantages of VPP is that it doesn't have geographical limits. There are two ways for controlling devices and transmitting information within a VPP: the centralized approach and the decentralized approach [1]. The centralized algorithms need a powerful central controller that collects the global information, processes the data, and requires a high bandwidth communication network. However, centralized controllers are costly and easy to suffer single-point failure. The algorithms need to be redesigned with the addition of new generators and loads. The centralized algorithms may not be accurate due to the unknown topology of the smart grid. The variety of configurations of power grid and communication network topology and the plug-and-play nature of the generating units make the topology of the network time-varying. Hence, a robust algorithm is required to operate correctly in the presence of limited and unreliable communication capabilities and also in the absence of centralized controllers.

There are different types of distributed optimization algorithms in the literature. The dual decomposition approach is one of the methods, and the consensus-based approach is the other one. Nowadays, many researchers have turned their attention toward consensus protocols to solve distributed smart grid problems. The consensus is effectively a network interaction protocol. Hence, the communication delays are unavoidable in consensus problems. In the stochastic environment, the consensus problem has been analyzed, along with communication losses and noises. The noises may be due to measurement errors, calculation errors, disturbance in the transmission medium, and quantization errors [4]. The main challenge is to reach a consensus with noise and communication delays in a switching communication network. The effect of communication delays and noise can be weakened by an appropriate monotonically decreasing gain function.

The VPP is an aggregation small decentralized generation. A decentralized energy management system (EMS) and its technical aspects are discussed in [5]. A distributed control strategy for VPPs has been proposed in [6]. The algorithm is tested in IEEE 34-node system without considering the voltage limits. A dynamic clustering algorithm has been applied to form VPP, which helps to improve voltage regulation and reduce power loss [7]. An energy management algorithm has been proposed considering microgrid as one of the units in the VPP [8]. A robust economic dispatch has been presented for a grid-connected VPP with energy storage (ES) [9]. The deep reinforcement learning approach is used to solve the linear nonconvex optimization problem.

A blockchain-based energy management platform has been developed in a residential VPP [10]. The users can interact with each other to trade energy for mutual benefits and provide network services, such as feed-in energy, reserve, and demand response, through the VPP. The short-term dynamic response has been improved through a coordinated frequency control of VPP [11]. The effect of communication delays of different communication networks with different bandwidths is also considered. The different communication infrastructures used for coordinated control distributed generations (DGs) in a VPP have been evaluated in [12]. The communication influences for the wireless technologies CDMA450 and LTE Advanced on the fully distributed optimization heuristic COHDA have been investigated. An interactive approach has been proposed for combined optimization of dynamic spectrum allocation and EV scheduling in the VPP to coordinate charging/discharging strategies of massive and dispersed EVs [13]. The proposed method has been developed under different imperfect communication scenarios. The VPP can provide some necessary information to help consumers improve their profits and trade with the electricity market. It has been proved that providing predictions can boost social total surplus stakeholders. A decentralized prediction provision algorithm in which consumers from each subregion only buy local predictions and exchange information with the VPP [14]. A distributed VPP dispatch algorithm is developed for the general VPP problem [15]. An optimal tracking problem (OTP) of electric vehicles via VPP has been solved using an alternating direction method of multipliers (ADMM)-based decentralized optimization algorithm [16]. With this algorithm, the network constraints that couple different EVs' charging power together are relaxed. The OTP model is thus decomposed into parallel single-EV charging subproblems that can be solved in a distributed manner.

In this work, a distributed consensus-based approach has been used for the optimal energy management of the VPP operation. The VPP operation is expressed in three modes in this chapter. The participants of VPP are assumed as agents and communicate among their neighbors. The algorithm is robust to communication delay and noise in the network. A consensus gain function, which is a function of the number of iterations and the maximum delay applied, is used to reduce noise and communication delay. The chapter is organized as follows. The basic concept of a VPP is described in Section 9.2. Section 9.3 explains the EMS of a VPP. Section 9.4 deals with the decentralized EMS. Section 9.5 formulates the problem. Section 9.6 introduces the consensus-based distributed algorithm. Section 9.7 describes the test systems considered in this study. Section 9.8 evaluates the simulation results. Finally, Section 9.9 concludes the work.

9.2 VIRTUAL POWER PLANT MODEL

9.2.1 VIRTUAL POWER PLANT CONCEPT

The VPP, a decentralized EMS, is a new concept. VPPs aggregate small generating units such as renewables, ES, and micro-CHP, enabling market participation for small generating units. According to the Fenix [17] project, the VPP is defined as: "A VPP aggregates the capacity of many diverse DERs, it creates a single operating profile from a composite of the parameters characterizing each DERs and can incorporate the impact of the network on aggregate DERs output. A VPP is a flexible representation of a portfolio of DERs that can be used to make contracts in the wholesale market and to offer services to the system operator [18]." VPP is similar to an autonomous microgrid. But the microgrid is represented by a single generator with a load, whereas VPP is a single generator.

Figure 9.1 shows a general overview of VPP. Similar to conventional power plants, VPP has its own operating characteristics such as scheduled output, ramp rates, generation limits, operating cost, voltage regulation capability, and reserve. VPP also incorporates controllable demands, parameters such as demand price elasticity, and load recovery patterns as characteristics. It may contain sources like photovoltaic, wind power, small hydrostations, biomass and biogas, electric vehicles [20], and combined heat and power generators. VPP is a good solution for controlling renewable energy sources (RES). When RES alone is connected to power networks, it might have some problems like lack of transmission capacity in the power network and variability of power output due to the variations of their primary energy sources. VPP may also contain small nonrenewable sources, controllable loads, and ES. The VPP function is explained in two configurations. The technical VPP (TVPP) and commercial VPP (CVPP) are the two different configurations of VPP. CVPP enables the VPP to participate in energy markets. The TVPP can perform CVPP's function and also provide balancing and ancillary services [2].

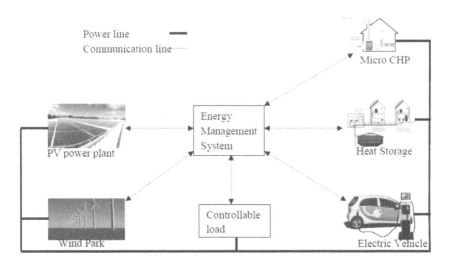

FIGURE 9.1 Overview of VPP [19].

9.3 ENERGY MANAGEMENT SYSTEM FOR VPP

The EMS is the heart of the VPP. EMS collects all the data from the generating units, controllable loads and storage, connected to the VPP. EMS performs some forecasting algorithms to provide bids in the market. According to the available data, EMS sends control signals to the elements in the VPP.

Figure 9.2 shows an EMS algorithm for VPP operation with an ES system (ESS) and controllable loads, where $E_s(t)$ represents the energy stored in the ESS. The centralized control center of VPP collects the demand and generation capacity data from all the participants of VPP and also gets the grid requirement. The main objective of the EMS algorithm is to maximize the profit of VPP. There are three different modes for a VPP operation: No Power Exchange mode, Power Export mode, and Power Import mode [19]. The EMS structure is the same in the three operating modes, whereas the load demand (D) is different. In the No Power Exchange mode, there will be No Power Exchange between the grid and VPP. In Power Export mode, the excess power available in VPP will be fed back to the grid. If there is a deficiency in power to supply grid requirements, this situation will be overcome by curtailing the controllable loads or with the help of ES. In Power Import mode, the power is injected from the grid to VPP. Usually, the Power Export mode operates in peak hours and the Power Import mode in off-peak hours to gain more profit.

9.4 DECENTRALIZED ENERGY MANAGEMENT SYSTEM

In this work, the VPP network is modeled as a multiagent system (MAS). Each node acts as an agent and has a local controller, which runs the local optimization algorithm and communicates information among the neighbors. Figure 9.3 shows the MAS model of a six-bus system. The local controller LC updates all the local variables as well as global variables and shares the global variables among its neighbors.

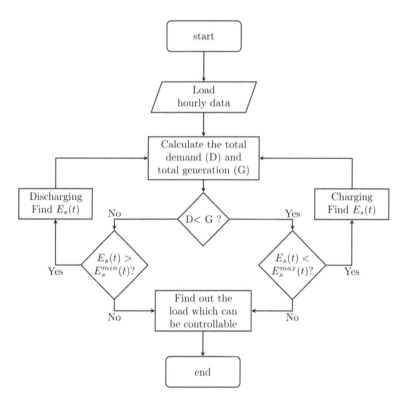

FIGURE 9.2 EMS algorithm for VPP operation [19].

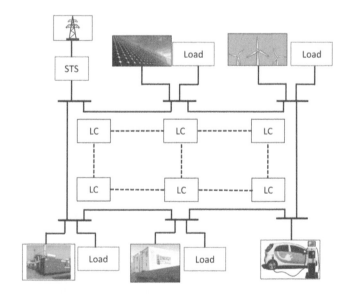

FIGURE 9.3 Communication network in a six-bus system.

Decentralized Energy Management System within VPP

The algorithm updates the global variable iteratively and finally converges to a common value.

The main objective of the optimal VPP operation is to maximize its profit considering power balance constraints, inequality constraints of generation, load and network constraints, etc. This optimization problem can be easily decomposed by relaxing the coupling constraints. The entire problem has been decomposed into subproblems and solved in the local controllers. All the participants of VPP try to maximize their individual profit leads to the maximization of profit of VPP. The main challenge is the local calculation of global variables. In this work, a consensus-based distributed approach is used, which is explained in Section 9.6. One of the main advantages of a distributed algorithm is that it preserves the privacy of the DERs. The participants do not share their private information with the control center, whereas they share their dispatch status with their neighbors within VPP. Since the algorithm is communication-based, the noise and communication delay has to be considered.

9.5 PROBLEM FORMULATION

A distributed optimal dispatch of VPP is considered in this work. The main objective of VPP optimal dispatch is to maximize its economic benefit by dispatching the outputs of renewable and nonrenewable generation units, the charging/discharging power of ES, and the amount of controllable loads [21]. It is an optimization problem with objective function as maximization of profit and is given by,

$$\max_{P_{dgn},\ P_{rn},P_{en},P_{cn},P_s} f = \sum_{n=1}^{N} \left\{ \begin{array}{l} \rho_d P_{dn} - C_{dgn}\left(P_{dgn}\right) - C_{rn}\left(P_{rn}\right) \\ -C_{en}\left(P_{en}\right) - C_{cn}\left(P_{cn}\right) \end{array} \right\} \\ -\rho_E P_s \tag{9.1}$$

where N is the number of nodes, ρ_E and ρ_s are the prices at which VPP buys electricity from the market and VPP sells electricity to its consumers, respectively. P_s is the amount of power injected into VPP, P_{dn} is the load at the nth node, which includes both interruptible and non-interruptible loads. The objective function has six terms. $\rho_d P_{dn}$ is the income from selling electricity to customers, $\rho_E P_s$ is the cost of purchasing electricity from the market, $C_{dgn}\left(P_{dgn}\right)$ is the fuel cost, $C_{rn}\left(P_{rn}\right)$ is the cost of renewable sources, $C_{en}\left(P_{en}\right)$ is the charging/discharging cost of ES devices, and $C_{cn}\left(P_{cn}\right)$ is the cost of interrupting loads. The above objective function is subjected to equality and inequality constraints. In this work, only DGs and interruptible loads are considered. Hence, the constraints which are related to DG and interruptible loads are discussed here.

    1. *Power balance constraints*:

The load within the VPP is served by generated outputs from the DG units, curtailing the controllable loads, and the remaining power is injected from the grid. Hence, this constraint is a coupling constraint. The power balance constraint is given by,

$$P_s = \sum_{i=1}^{N}\left(P_{dn} - P_{dgn} - P_{cn}\right) \tag{9.2}$$

2. *DG constraints*:

This work mainly focuses on a particular time block. Hence, only the capacity constraint is considered and is given by,

$$P_{dgn}^{min} \leq P_{dgn} \leq P_{dgn}^{max}, \; n = 1,2,....N \tag{9.3}$$

where P_{dgn}^{min} and P_{dgn}^{max} are the lower and upper limits of the generation limits of the nth generator.

3. *Interruptible load constraints*:

$$0 \leq P_{cn} \leq P_{cn}^{max}, \; n = 1,2,....N \tag{9.4}$$

where P_{cn}^{max} is the maximum amount of interruptible load at node n.

4. *Transmission line constraints*:

The transmission line constraints obtained from DC power flow is given by,

$$-T_l \leq \sum_{n=1}^{N}\eta_{ln}\left(P_{dgn} + P_{cn} - P_{dn}\right) \leq T_l \; l = 1,2,....L \tag{9.5}$$

where, T_l is the transmission limit of line l, η_{ln} is the sensitivity of the power injected at node n with respect to the power flow of line l, and L is the total number of transmission lines in the system.

9.6 CONSENSUS-BASED DISTRIBUTED ALGORITHM

A consensus-based approach is used to solve the optimal dispatch of VPP. The algorithm is based on a first-order discrete-time consensus protocol, which is similar to the proportional controller of the control system. A minor modification is included to the protocol to suppress the effect of noise and communication delay.

9.6.1 THE FIRST-ORDER DISCRETE-TIME CONSENSUS PROTOCOL

Let $x_i \in \mathfrak{R}$, is the state variable of node i, can be any of the physical quantities like power mismatch, incremental cost, and output power. The system reaches consensus when all the nodes attain the same state, i.e., $x_i = x_j \; \forall \; (i,j)$. Considering all the nodes have first-order dynamics, a standard linear consensus protocol [22] is given by,

$$x_i(t) = \sum_{j}a_{ij}\left(x_j - x_i\right) \tag{9.6}$$

Decentralized Energy Management System within VPP

If a group of agents is following the protocol in Eq. (9.6), the collective dynamics is given by,

$$x_i = -Lx \qquad (9.7)$$

where L is the Laplacian matrix of the graph. A discrete-time linear consensus protocol [22] having first-order dynamics can be represented as,

$$x_i(k+1) - x_i(k) = u_i(k) \Rightarrow$$
$$x_i(k+1) = x_i(k) + u_i(k) \qquad (9.8)$$

where, $u_i(k)$ depends on the information state of the neighbors of ith node and is given by,

$$u_i(k) = \sum_j a_{ij}\left(x_j(k) - x_i(k)\right) \qquad (9.9)$$

From Eqs. (9.8) and (9.9), the discrete-time consensus algorithm can be written as,

$$x_i(k+1) = \sum_{j=1}^{N} d_{ij} x_j(k) \qquad (9.10)$$

where k denotes the discrete-time index; d_{ij} is the (i,j)th element of the row-stochastic matrix D, which can be defined as,

$$d_{ij} = \frac{l_{ij}}{\sum_{j=1}^{n}|l_{ij}|} \qquad (9.11)$$

From Eq. (9.10), it is clear that the current state of each state variable depends on the previous state of the other state variables (state variables of the neighboring nodes). In this work, a minor modification is applied to this consensus protocol to reduce the effects of noise and communication delay.

The consensus protocol considering noise and communication delay is given by,

$$x_i(k+1) = x_i(k) + c(k)\sum_{j\in N_i} d_{ij}\left(x_j(k-\tau_{ij}) + \sigma_{ij} - x_i(k)\right) \qquad (9.12)$$

where $x_i(k+1)$ and $x_i(k)$ are the consensus variables at node i at $(k+1)$th and kth iteration, respectively, $c(k)$ is the consensus gain function, N_i is the no. of neighbors of node i, τ_{ij} is the communication delay, σ_{ij} is the noise, and d_{ij} is the (i, j)th element of the doubly stochastic matrix D as given in Eq. (9.13). The d_{ij} is formed by improved metropolis method [23] and is given by,

$$d_{ij} = \begin{cases} \dfrac{1}{\left(\max\left(d_i,d_j\right)+1\right)}, & j \in N_i \\[2ex] 1 - \displaystyle\sum_{j \in N_i} \dfrac{1}{\left(\max\left(d_i,d_j\right)+1\right)}, & i = j \\[2ex] 0, & \textit{otherwise} \end{cases} \tag{9.13}$$

where d_i and d_j are the degrees of nodes i and j, respectively. The consensus gain function is a monotonically decreasing function which satisfies the conditions [24],

$$\sum_{k=0}^{\infty} c(k) = +\infty$$
$$\sum_{k=0}^{\infty} c(k) < +\infty \tag{9.14}$$

The consensus gain function considered in this work is given by,

$$c(k) = \frac{\bar{d}}{\sqrt{k + \bar{d}}} \tag{9.15}$$

where k is the iteration index and $\bar{d}$ is the maximum delay applied.

In this work, the distributed optimal dispatch of VPP is solved using two consensus protocols in parallel $-\lambda$ consensus and ΔP consensus. The flow chart of the algorithm is shown in Figure 9.4. The consensus variables' incremental cost and power imbalance update in each iteration through the consensus protocol given in Eq. (9.12). From the incremental cost, the power generation P_{dgi} and the curtailable load P_{ci} are calculated. At optimal condition, the incremental cost λ converges to a constant value λ_{opt}, and the power imbalance converges to zero at all buses. Moreover, this λ_{opt} may not be the same as the locational marginal price of the buses.

9.7 CASE STUDY

In this section, two different VPP systems are studied. At first, a small VPP, shown in Figure 9.5, is used to explain the centralized energy management algorithm, where the VPP contains two DGs and two dynamic loads. A centralized controller CC controls the operation of the VPP through the EMS algorithm. Figure 9.5 shows that all the elements of VPP communicate with the central controller, which sends control signals to the corresponding participants. Here, VPP is connected to a radial distribution network.

Then, the distributed algorithm is evaluated using a larger VPP. A 15-node VPP is considered, as shown in Figure 9.6. It contains 8 DG units and two controllable loads. The loads connected at nodes 4 and 7 can be curtailed up to 30 and 40 kW, respectively. The VPP is connected to the main grid at node 1, and the maximum

Decentralized Energy Management System within VPP

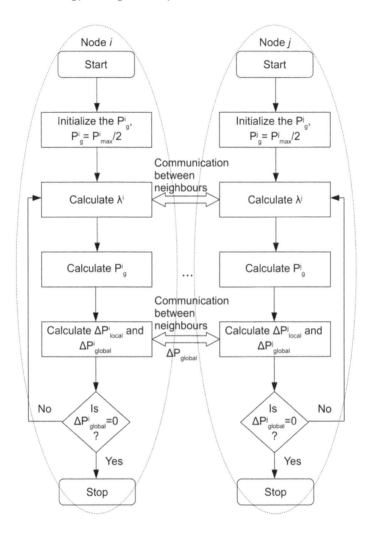

FIGURE 9.4 Flow chart of consensus-based distributed algorithm [25].

exchange capacity is assumed to be 1200 kW. The VPP injects power into the grid when the energy market price is greater than the production cost of VPP, and the VPP imports power when the VPP production cost is greater than the energy market price.

9.8 SIMULATION RESULTS AND DISCUSSIONS

9.8.1 Evaluating Small VPP, VPP1

A small VPP model is implemented using Matlab-Simulink. The schematic diagram of the main grid and VPP is shown in Figure 9.5 [19]. The DG is modeled as a synchronous machine with a hydraulic turbine governor and an excitation model. It is

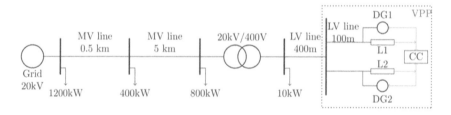

FIGURE 9.5 The schematic diagram of VPP and the main grid [19].

a 31.3-kVA, 50-Hz, 1500-rpm, 400-V synchronous generator. The maximum active output power of each DG is assumed to be 30 kW. Any renewable generation can replace the synchronous machine. The loads L1 and L2 are modeled as dynamic loads with a maximum demand of 31.3 kW. Any domestic load can replace the dynamic load. The VPP with two DGs and two loads are connected to the main grid.

The VPP operation during its three modes is analyzed using three different scenarios. Figure 9.7(a) and (b) shows the power exchange of VPP with the main grid and the load generation balance of VPP, respectively. There is No Power Exchange between the grid and VPP in the first scenario (0–25 s). The mismatch between the demand and generation is due to losses. The power generated within VPP is consumed by its own loads. The excess power within the VPP is injected into the grid in the second scenario (25–75 s). This mode is likely to happen in peak hours. The load increases again in the third scenario (75–100 s), and the grid supplies the deficit power. The simulation results are tabulated in Table 9.1.

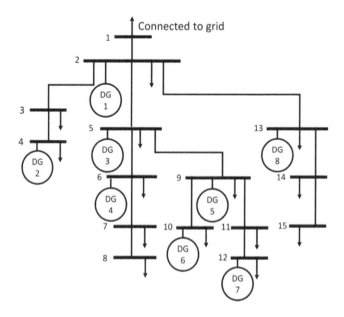

FIGURE 9.6 Single line diagram of a 15-node VPP.

Decentralized Energy Management System within VPP

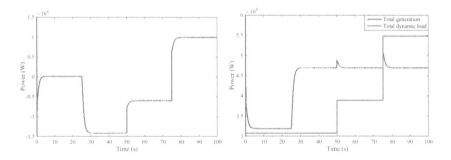

FIGURE 9.7 (a) Power exchange between VPP and grid and (b) total load and generation within VPP.

9.8.2 Evaluating Large VPP, VPP2

The performance of the distributed algorithm is assessed using different case studies on a 15-node VPP [26]. The simulations are performed in MATLAB. As shown in Figure 9.6, the VPP2 contains eight DGs and two controllable loads. Each node has a local controller, where the local optimization happens and communicates the updated consensus variables among its neighboring nodes. The load data and the quadratic cost characteristics used for DGs are depicted in Tables 9.2 and 9.3. The costs of curtailing load paid to consumers are calculated by $C(P) = 0.01P^2 + 3P$ for node 4 and $C(P) = 0.01P^2 + 1.5P$ for node 7, respectively, where P is the load unserved.

A dynamically switching communication network is considered in all the case studies. The price of the electricity at which the VPP sells power to consumers is assumed to be the same as the optimum incremental cost.

9.8.2.1 Case Study I – No Power Exchange Between VPP and the Grid

The No Power Exchange mode of VPP is analyzed in this case study. The load demand of the system is assumed to be below the total generation capacity of VPP. The total load served in this case study is 704 kW. All the loads are served by the

TABLE 9.1
Different Modes of Operation

Duration (s)	VPP Operation Mode	Power Generated in VPP	Power Exchange with the Grid
0–25	No Power Exchange mode	(Total load + losses) within VPP	No Power Exchange with the grid
25–75	Power Export mode	(Total load + losses) in VPP + power supplied to the grid	Excess power is injected into the grid
75–100	Power Import mode	(Total load + losses) in VPP – power supplied from the grid	Deficit power is injected from the grid

TABLE 9.2
The Load Data of 15-Node VPP

Node	1	2	3	4	5	6	7	8	9	10	11	12	13	14	15
Load (kW)	0	33	50	50	75	58	58	58	42	33	33	100	42	42	30

TABLE 9.3
Generator Cost Characteristics of VPP

Sl. No.	Node	a ($/kW²)	b ($/kW)	P_{dg}^{min}	P_{dg}^{max}
1	2	0.01	10.5	20	85
2	4	0.01	6.5	35	115
3	5	0.01	9.2	30	110
4	6	0.01	12.6	20	95
5	9	0.01	7.2	25	80
6	10	0.01	7	30	90
7	12	0.01	10.1	30	105
8	13	0.01	12.7	20	90

DGs present in the VPP. The algorithm converges fast, as shown in Figure 9.8(a). The incremental cost at every node converges to λ_{optim}, 13.84 Monetary units/kW. Figure 9.8(b) shows the output powers of individual DG units. The total profit obtained is 2668.1 Monetary units.

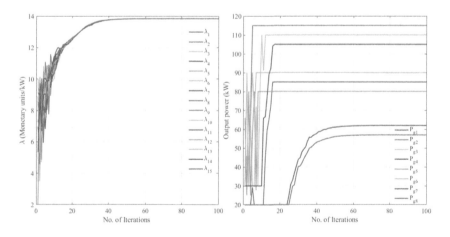

FIGURE 9.8 (a) Incremental cost and (b) DG power outputs.

9.8.2.2 Case Study II – No Power Exchange Between VPP and the Grid Considering Controllable Loads

In this case study, the No Power Exchange mode is considered. But the excess load is served by curtailing the interruptible loads. Figure 9.9 shows the algorithm converges fast. The incremental costs of all the nodes converge to the optimum value of 14.22 Monetary units/kW, as shown in Figure 9.9(a). The total served load, unserved load, and profit are 742 kW, 70 kW, and 2767.8 Monetary units.

Figure 9.9(b) shows the power outputs of individual DGs in the system. There is a slight increase in the profit and the optimum incremental cost since the load served in this case study is more than case I.

9.8.2.3 Case Study III – Power Export Mode of VPP

The Power export mode of VPP is studied in this case study. A considerable amount of power is injected into the grid at the cost of $(\lambda_{optim} + 2)$ Monetary units. Figure 9.10(a) shows the faster convergence of the algorithm. The optimum incremental cost is obtained as 14.44 Monetary units/kW. The total load served within the VPP is 634 kW, whereas the unserved load is 70 kW, and 130 kW power is injected into the grid. The total profit obtained is 3193.5 Monetary units.

The output powers of DG units are shown in Figure 9.10(b). There is a significant improvement in the profit than the above case studies since the power is exported to the grid at a higher price. This is likely to happen during peak hours.

9.8.2.4 Case Study IV – Power Import Mode of VPP

The power import mode of VPP is considered in this case study, which is more often in off-peak hours. The VPP imports power from the grid since the electricity price is lower in the energy market than the production cost of VPP. The price of electricity at the energy market is assumed to be $(\lambda_{optim} - 4)$ Monetary units/kW. The interruptible loads are not curtailing in this case study. The incremental cost

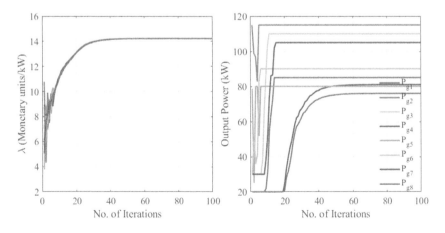

FIGURE 9.9 (a) Incremental cost and (b) DG power outputs.

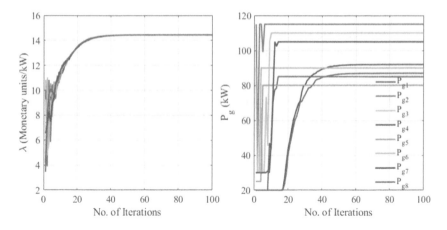

FIGURE 9.10 (a) Incremental cost and (b) DG power outputs.

at each node converges to the optimum value, as shown in Figure 9.11(a), which is obtained as 13.96 Monetary units/kW. The total load served within VPP is 846 kW; out of this, 130 kW is injected from the grid. The total profit is obtained as 3273.3 Monetary units.

The output powers of DGs are shown in Figure 9.11(b). There is a significant increase in profit than the first two case studies since a large load is served at a lower price.

9.8.2.5 Case Study V – VPP Operation with Communication Delay and Noise

The effect of communication delay and noise is analyzed in this case study. The noise is modeled as a uniformly distributed function over $[-N_A\ N_A]$, where N_A is the

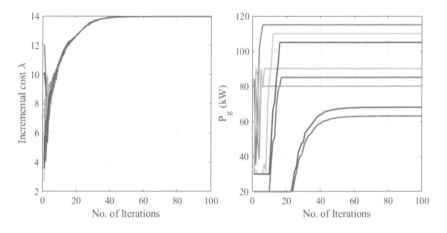

FIGURE 9.11 (a) Incremental cost and (b) DG power outputs.

Decentralized Energy Management System within VPP

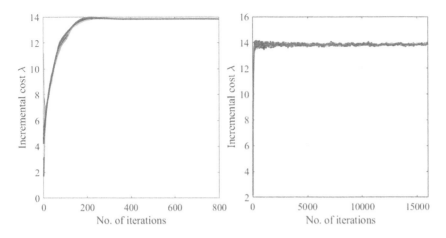

FIGURE 9.12 Incremental cost (a) $N_A = 0.01$ and (b) $N_A = 10$.

noise amplitude. The maximum noise amplitude is assumed to be 10% of the optimum value. The consensus gain function used for simulation is given by,

$$c(k) = \frac{2}{\sqrt{k+2}}.$$

The simulation results of different noise scenarios are shown in Figure 9.12. The consensus gain function could effectively suppress the effect of noise. Nevertheless, the performance of the algorithm degrades with an increase in noise amplitude, as shown in Table 9.4.

The communication delay is represented in terms of no. of iterations. Table 9.5 shows the different communication delay scenarios, where τ represents the communication delay in number of iterations and ρ represents its corresponding probability. $\tau = 0$ indicates there is no delay. $\tau = 1$ and $\tau = 2$ mean the signal is delayed by one iteration and two iterations, respectively. The simulation is performed considering the noise scenario IV as given in Table 9.4. The algorithm could easily handle the communication delay, as shown in Table 9.5.

TABLE 9.4
Statistical Results Obtained in 100 Runs for Different Noise Scenarios

Sl. No.	Noise Amplitude	k_{min}	k_{max}	k_{avg}
1	$N_A = 10$	6826	25,003	17,599
2	$N_A = 1$	4270	23,645	12,703
3	$N_A = 0.1$	1218	4745	2974
4	$N_A = 0.01$	739	944	805

TABLE 9.5

Statistical Results Obtained in 100 Runs for Different Communication Delay Scenarios

Sl. No.	$\rho\ (\tau = 0)$	$\rho\ (\tau = 1)$	$\rho\ (\tau = 2)$	k_{min}	k_{max}	k_{avg}
1	1	0	0	739	944	804
2	0.7	0.15	0.15	781	977	842
3	0.5	0.25	0.25	787	989	854
4	0.3	0.35	0.35	799	1005	870
5	0	0.5	0.5	825	1070	904

9.9 CONCLUSION

The distributed optimal dispatch of VPP by using a consensus-based algorithm has been proposed. The proposed algorithm uses a first-order discrete-time consensus protocol for both incremental cost consensus and ΔP consensus. The consensus protocol is similar to the proportional controller, which reduces the deviation in each iteration. This algorithm suppresses the malignant effects of communication delay and noise through a monotonically decreasing consensus gain. This property of the algorithm improves the robustness of the algorithm. One of the main advantages of this algorithm is that the generating units need not share their cost characteristics, thereby preserving their private information.

REFERENCES

1. Y. Chen, T. Li, C. Zhao, and W. Wei, "Decentralized provision of renewable predictions within a virtual power plant", IEEE Transactions on Power Systems, vol. 36, no. 3, pp. 2652–2662, May 2021.
2. H. Saboori, M. Mohammadi, and R. Taghe, "Virtual power plant (VPP), definition, concept, components and types," *2011 Asia-Pacific Power and Energy Engineering Conference*, 2011, pp. 1–4.
3. K. E. Bakari and W. L. Kling, "Virtual power plants: An answer to increasing distributed generation," *2010 IEEE PES Innovative Smart Grid Technologies Conference Europe (ISGT Europe)*, 2010, pp. 1–6.
4. S. Liu, L. Xie, and H. Zhang, "Distributed consensus for multi-agent systems with delays and noises in transmission channels," Automatica, vol. 47, no. 5, pp. 920–934, 2011.
5. T. G. Werner and R. Remberg, "Technical, economical and regulatory aspects of virtual power plants," *2008 Third International Conference on Electric Utility Deregulation and Restructuring and Power Technologies*, 2008, pp. 2427–2433.
6. W. Zheng, W. Wu, B. Zhang, Z. Li, and Y. Liu, "Fully distributed multi-area economic dispatch method for active distribution networks," IET Generation, Transmission & Distribution, vol. 9, no. 12, pp. 1341–1351, 2015.

7. R. Zhang, B. Hredzak, and J. Fletcher, "Dynamic aggregation of energy storage systems into virtual power plants using distributed real-time clustering algorithm," IEEE Transactions on Industrial Electronics, vol. 68, no. 11, pp. 11002–11013, 2021.

8. S. I. Taheri, M. B. Salles, and E. C. Costa, "Optimal cost management of distributed generation units and microgrids for virtual power plant scheduling," IEEE Access, vol. 8, pp. 208449–208461, 2020.

9. D. Fang, X. Guan, B. Hu, Y. Peng, M. Chen, and K. Hwang, "Deep reinforcement learning for scenario based robust economic dispatch strategy in internet of energy," IEEE Internet of Things Journal, vol. 8, no. 12, pp. 9654–9663, 2021.

10. Q. Yang, H. Wang, T. Wang, S. Zhang, X. Wu, and H. Wang, "Blockchain-based decentralized energy management platform for residential distributed energy resources in a virtual power plant," Applied Energy, vol. 294, p. 117026, 2021.

11. W. Zhong, J. Chen, M. Liu, M. A. A. Murad, and F. Milano, "Coordinated control of virtual power plants to improve power system short-term dynamics," Energies, vol. 14, no. 4, p. 1182, 2021.

12. F. Oest, M. Radtke, M. Blank-Babazadeh, S. Holly, and S. Lehnhoff, "Evaluation of communication infrastructures for distributed optimization of virtual power plant schedules," Energies, vol. 14, no. 5, p. 1226, 2021.

13. B. Zhou, K. Zhang, K. W. Chan, C. Li, X. Lu, S. Bu, and X. Gao, "Optimal coordination of electric vehicles for virtual power plants with dynamic communication spectrum allocation," IEEE Transactions on Industrial Informatics, vol. 17, no. 1, pp. 450–462, 2020.

14. Y. Chen, T. Li, C. Zhao, and W. Wei, "Decentralized provision of renewable predictions within a virtual power plant," IEEE Transactions on Power Systems, vol. 36, no. 3, pp. 2652–2662, 2020.

15. G. Chen and J. Li, "A fully distributed ADMM-based dispatch approach for virtual power plant problems," Applied Mathematical Modelling, vol. 58, pp. 300–312, 2018.

16. Z. Li, Q. Guo, H. Sun, and H. Su, "ADMM-based decentralized demand response method in electric vehicle virtual power plant," in *2016 IEEE Power and Energy Society General Meeting (PESGM)*. IEEE, 2016, pp. 1–5.

17. C. Kieny, B. Berseneff, N. Hadjsaid, Y. Besanger and J. Maire, "On the concept and the interest of virtual power plant: Some results from the European project Fenix," *2009 IEEE Power & Energy Society General Meeting*, 2009, pp. 1–6.

18. D. Pudjianto, C. Ramsay, and G. Strbac, "Virtual power plant and system integration of distributed energy resources," IET Renewable Power Generation, vol. 1, no. 1, pp. 10–16, 2007.

19. P. M. Naina, H. Rajamani, and K. S. Swarup, "Modeling and simulation of virtual power plant in energy management system applications", In *2017 7th International Conference on Power Systems (ICPS)*, pp. 392–397. IEEE, 2017.

20. C. Binding et al., "Electric vehicle fleet integration in the Danish EDISON project – A virtual power plant on the island of Bornholm," *IEEE PES General Meeting*, 2010, pp. 1–8.

21. E. Mashhour and S. M. Moghaddas-Tafreshi, "Bidding strategy of virtual power plant for participating in energy and spinning reserve markets—Part I: Problem formulation," IEEE Transactions on Power Systems, vol. 26, no. 2, pp. 949–956, May 2011.

22. R. Olfati-Saber, J. A. Fax, and R. M. Murray, "Consensus and cooperation in networked multi-agent systems", Proceedings of the IEEE, vol. 95, no. 1, pp. 215–233, 2007.

23. Y. Xu and W. Liu, "Novel multiagent based load restoration algorithm for microgrids," IEEE Transactions on Smart Grid, vol. 2, no. 1, pp. 152–161, 2011.

24. S. Liu, L. Xie, and H. Zhang, "Distributed consensus for multi-agent systems with delays and noises in transmission channels," Automatica, vol. 47, no. 5, pp. 920–934, 2011.
25. P. M. Naina and K. S. Swarup, "Double consensus based optimal dispatch considering communication delay and noise," *2020 21st National Power Systems Conference (NPSC)*, 2020, pp. 1–6.
26. E. Mashhour and S. M. Moghaddas-Tafreshi, "Bidding strategy of virtual power plant for participating in energy and spinning reserve markets—Part II: Numerical analysis," IEEE Transactions on Power Systems, vol. 26, no. 2, pp. 957–964, May 2011.

10 Distributed Synchronism System Based on TSN and PTP for Virtual Power Plant

V. Pallares-Lopez, I. M. Moreno-Garcia,
R. Real-Calvo, V. Arenas-Ramos,
M. Gonzalez-Redondo, and I. Santiago
Departamento Ingeniería Electrónica y de Computadores,
Universidad de Córdoba, Córdoba, Spain

CONTENTS

10.1 Introduction 191
10.2 Application-Dependent Distributed Synchronism Levels
(NTP, PTP, and TSN) 195
10.3 Real-Time Embedded Systems For Distributed Control in the
New VPP Framework and Fast Disturbance Detection............................ 197
10.4 Real-Time PTP Architecture For Distributed Generation System
Synchronisation 198
10.5 Architecture With MPMU and TSN Standard As Key
Components in the Future of VPPs 200
10.6 The IEC 61850 Standard As a Reference Framework For VPPs.................202
10.7 New Architecture For Capturing, Processing, and
Managing Edge Databases 203
References..204

10.1 INTRODUCTION

Classical energy production models are being replaced by distributed energy resource (DER) models that involve customers as producers and consumers of energy, prosumers. Grid-edge technologies [1] are facilitating this transition with proximity energy generation, to the detriment of energy produced in conventional power plants. This new model makes it possible to reduce losses associated with transmission and distribution lines, promoting microgeneration with the participation of the entire community.

DOI: 10.1201/9781003257202-11

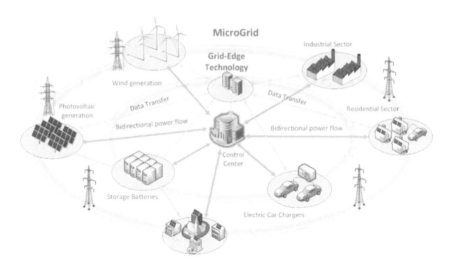

FIGURE 10.1 Microgrid basic model.

Figure 10.1 shows a schematic of a microgrid with the basic elements that compose it: energy generation sources, such as homes or buildings with production and storage capacity; supply grid, which allows a bidirectional energy flow to favour energy storage and exchange between customers; a common space for photovoltaic or wind energy generation; and a collective storage system.

In this context, real-time control of the electricity grid with grid-edge technology [1] will allow the management of energy flow generated for consumption or for storage in the collective use batteries. The control system must generate messages in a matter of milliseconds to request direct consumption at the moment of generation or for use of the reserved energy, so that the control system can switch the battery from charging mode to power supply mode. The first is to guarantee energy supply to consumers in the event of a contingency, optimising their consumption in an intelligent way according to the cost of the defined sections throughout the day. The second objective is to provide the distribution company with the necessary time to increase the central energy resources before the local system requires it.

In this context, this energy transfer must be precisely managed from a central control system. As mentioned above, if the microgrid has common storage systems, energy can be redirected to be stored and consumed at times of lower production. For this microgeneration to be managed as a single-generation system, precise synchronisation, monitoring, and control are required [2, 3]. This chapter takes a closer look at the contribution of the IEEE 1588 v2 [4] or Precision Time Protocol (PTP) and IEEE 802.1AS [5] or Time-Sensitive Networking (TSN) standards to this area. The latter is a clear evolution of the former and allows synchronisation levels of over 1 μs to be achieved.

The three key points of a microgrid architecture and by extension virtual power plant (VPP) architecture are the use of grid-edge technology [1], which encompasses smart technology connected to the Grid-Edge, high-performance peer-to-peer (P2P)

Distributed Synchronism System Based on TSN and PTP

communications for local autonomy, and increased and improved data management with edge-computing techniques. A tiered architecture is used to integrate the control of microgrids and its real-time data bus with management and analysis in the cloud [6]. In this architecture, it is essential to ensure high-precision synchronisation, so that the equipment involved is coordinated for monitoring, and the control of the protection systems meets the requirements of various international standards. Therefore, the choice of protocol (PTP) or (TSN) [2] will depend on the level of requirements of the protection standards for critical event detection.

A P2P network is a decentralised communication architecture, where agents, called peers, can act as both client and server simultaneously. The main advantage of these networks is their high scalability since the exchange of information takes place directly between end-users without going through an intermediate server.

The concept of VPP [7–9] is broader because it proposes to group these microgrids in residential, commercial, or industrial environments to control them jointly under a type of pricing or DER programmes. In addition, it can group different microgrids from residential or industrial sectors located in nearby sectors (Figure 10.2). It is a type of collaborative grouping among communities to share different generation and storage systems. Local energy resources are shared, avoiding losses in transport; resources are optimised based on the consumption needs of the community, and energy savings are encouraged as they feel part of the process.

For a virtual management system, such as VPPs, there must be a supervisory system operating at a higher level that coordinates all data from the communications standards commonly used in the energy sector [10]. This communication standard should centrally control the distributed monitoring and control equipment in each sector. Grid-edge technology contributes favourably to this type of decentralised processing architecture. Local devices monitor, process, store, and generate events on local servers.

The Open Field Message Bus (OpenFMB) standard is a framework and reference architecture that enables the interoperability of these distributed energy systems and can contribute to the development of advanced distribution management systems (ADMSs). Some researchers [11] propose the joint use of ADMS and remote hardware-in-the-loop (rHIL) for the validation of systems in the laboratory before the development of field tests. It uses the OpenFMB standard as a reference for the joint management of data from smart inverters, smart meters, phasor measurement units (PMUs), etc.

In terms of measurement synchronisation, the PTP protocol for smart grids [12] guarantees a distributed synchronism of up to 1 ms, and the IEC 61850 standard [13] guarantees transmission in real time, according to a very widespread communication model.

In the model architecture shown in Figure 10.2, at least one μPMU is included for each microgrid acting as a phasor meter to extend the synchronism to the rest of the measurement systems. This reference will mark the beginning of the sampling process of the acquisition systems, which is essential to establish relationships among the events taking place at the power generation points. In a real scenario, the μPMUs would be located at grid injection points, monitoring production data and reporting on grid stability.

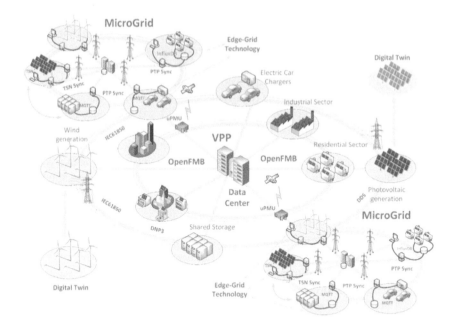

FIGURE 10.2 VPP model for the residential or industrial sector.

In short, a VPP will need to exercise precise control for the guaranteed management of energy consumption, and it is vital to guarantee a high level of synchronisation for the capture, transmission, and storage of all electrical parameters. The use of a model based on the IEC 61850 standard guarantees a high degree of homogeneity for the exchange of information between a central system and the distributed generation systems. Likewise, the incorporation of high-performance synchronism protocols such as PTP or TSN guarantees the correlation of the data captured at various points in the network.

IEEE 1815-2010 [14] or Distributed Network Protocol, version 3 (DNP3) is the most widely used standard in the US energy market. However, the European standard IEC 61850 is gaining widespread recognition as a benchmark for all energy management systems worldwide. Many companies currently using DNP3 are choosing to adopt the cross-cutting functionality of both DNP3 and IEC 61850. The main difference between DNP3 and IEC 61850 is that the IEC standard focuses on the context of the data, whereas DNP3 focuses on the data and largely leaves contextualisation to be handled by engineers.

The IEC 61850 protocol, therefore, integrates context into the system by mapping data to logical nodes with predefined names. This ensures that context is never lost amidst the confusion of data collection. This is one of the reasons why it is used by many researchers as a reference for the smart grids of the future [13, 15].

On the other hand, in the architecture of Figure 10.2, the possible existence of digital twins for wind and PV plants has been included as an example. This is a natural development in the context of current research into the evolution of smart grids.

As a reference to the importance of this technology, we can cite the initiatives of the company Opal-RT Technologies [16]. The tools developed by Opal-RT allow the development of twins for the energy sector with hardware-in-loop (HIL) technology. This set of tools allows the synthesis at the laboratory level of physical models that respond in real time like real models. An example from this discipline is the work published by Wang et al. in 2021 [17].

10.2 APPLICATION-DEPENDENT DISTRIBUTED SYNCHRONISM LEVELS (NTP, PTP, AND TSN)

By means of any of the existing Network Time Protocol (NTP) servers based on Universal Coordinated Time (UTC), it is possible to synchronise the clocks of the monitoring system. Currently, synchronisation levels of the order of 10 ms can be achieved using the NTPv4 version. On the other hand, uncertainty levels lower than 1 ms are reached through the PTPv2 protocol. This is especially intended for distributed synchronisation in industrial applications. In a 2013 publication [12], the TIC240 research group at the University of Cordoba analysed the PTPv2 standard under favourable technological conditions and observed that uncertainties of less than 50 µs can be achieved. However, even with these improvements, these levels of uncertainty are not adequate to coordinate the simultaneous capture in distributed acquisition systems that allow the distributed control of microgrids with DC generation. On the other hand, the same research group has experimentally analysed the TSN standard in different working conditions for simultaneous and distributed capture [3]. In their research, they have detected uncertainty levels below 100 ns, which would allow for optimal coordinated control of inverters.

Service providers are slowly modernising their transmission infrastructure, using standards such as TSN [5] to provide sub-millisecond synchronised measurements in real-time phase, frequency, and voltage measurements. Instead of the traditional method of tracking the alternating signal, transmission over the Ethernet network is used to share real-time measurements. This methodology will make it possible to create a virtual synchronisation and approach the synchronisation problem from a different perspective. In this sense, the aforementioned TSN network standard [5], allows the transmission of critical data in real time with a conventional Ethernet infrastructure.

The technology used by different authors to analyse the quality of the TSN standard corroborates the above. Particularly relevant are the contributions of Agarwal in 2019, as a member of the IEEE and with the support of Lamar University (Texas) and the multinational company National Instruments [18]. The study analysed how the TSN protocol allows the synchronisation of the data flowing through the network with a common time reference for all the measurement devices used in the network, allowing a reduction in the cost of the systems, greater security, and reliability, as well as synchronisation without the need to use GPS systems inserted in each of the measurement points.

This study proposes the design and construction of a real-time simulation prototype of a software library for the implementation of the TSN protocol, considering various scenarios at multiple prioritisation levels, allowing the control and monitoring of data for the configuration, scheduling, and prioritisation of data traffic. The implementation of the simulation prototype was performed using a real-time innovation (RTI) connectivity framework, the Data Distribution Service (DDS) standard [19, 20], and MATLAB/SIMULINK libraries. A new approach for TSN-based real-time multilevel communications using DDS for synchronised measurement data transfer in three-phase power systems is presented. Results showed that the prototype achieves low latency and high-throughput data transmissions, making it a desirable option in communication systems where potentially critical data is involved, such as in microgrids, smart cities, and military applications [18].

In [21], Smith studies the latency guarantee and synchronisation accuracy characteristics of TSN standards in the context of power systems and discusses their potential applications in the Micro-Network domain. In this study, the performance of TSN latency guarantees is compared in a network with varying traffic load with common UDP and TCP protocols. The main objective is to show that even with high traffic levels in the communication network, the latency guarantee features of TSN will allow fast data transmission through the network, where non-TSN traffic would experience longer delays. In addition, the synchronisation performance of the 802.1AS standard is analysed as a potential means of generating synchronised signals that contribute to the synchronisation of distributed resources in a microgrid. From the tests performed, it is worth noting that better latency levels were obtained for any level of network traffic load with TSN compared to the equivalent tests performed with TCP and UDP protocols. The tests performed with TSN in this study achieved latency results in the order of 2 ms. Although it was possible to reduce these levels to 1 ms by altering the CNC period set point, this led to an undesirable increase in the computational requirements of CompactRIO (cRIO).

Future synchronisation systems for VPPs will rely on PTP and TSN to maintain a single, high-quality time reference frame. This synchronism technology can be applied to new infrastructures dedicated to the monitoring of microgeneration power grids and can be applied to larger areas with the involvement of μPMU. The proposed infrastructure works with standard Ethernet networks, so that measurement equipment can coexist with other data transmission systems in various joint production and consumption scenarios. In practice, this means that standard Ethernet functionality is extended, so that message latency is guaranteed with critical and non-critical traffic. This is to ensure that real-time control can be extended outside the area of local operations.

From a technological perspective, the combined use of PTP and TSN standards is proposed as a novelty to guarantee the precise synchronisation of all production, consumption, and energy injection data. The choice of one standard or the other will depend on the critical levels of data according to the specifications of the IEC 61850 communications standard [13], which establishes two levels of requirements for packet transfer depending on the application. For electrical parameters, it uses a type of message with less demanding response times, and for events it uses other types of messages that guarantee a timely response by the requirements of the safety standards, to allow the rapid intervention of the protection relays.

10.3 REAL-TIME EMBEDDED SYSTEMS FOR DISTRIBUTED CONTROL IN THE NEW VPP FRAMEWORK AND FAST DISTURBANCE DETECTION

To address the challenge of the increasing and progressive integration of DERs, several multinational high-tech companies [22], such as Wipro, National Instruments, RTI, and Cisco, joined forces to work on common projects to initiate trials with Micro-Networks, under the Industrial Internet Consortium (IIC) [23]. The objective is to evaluate and develop 100% renewable energy generation techniques with synchronisation and communications based on the Ethernet standard.

This challenge can be met with the participation of embedded systems working in real time. The companies mentioned above include among their products devices managed by real-time operating systems (RTOSs) such as VwWork or LinuxRT for the execution of tasks with a priority order according to the response time specifications of international standards. When addressing the control of microgrids, generally all those actions that allow to achieve voltage and frequency regulation, to achieve a balance of active and reactive power between DERs, and to allow the reconnection and resynchronisation of the main grid acting as an anti-Islanding system are addressed (Figure 10.3).

In the article [24], a real-time system is designed with the RTOS VxWorks for event detection according to the time specifications of the main international protection standards or more recently in [25] with μPMUs as the main technological reference for transient event detection.

Systems with RTOS can perform these tasks and others that are of high priority such as voltage and phase angle synchronisation of the microgrids isolated from the main grid [24] for successful reconnection to the grid. The phase angles of the two systems must be closely matched. To avoid power loss, the microgrid controllers use phase loops and special control methods to synchronise the frequency and phase of the supply voltage with the mains [6].

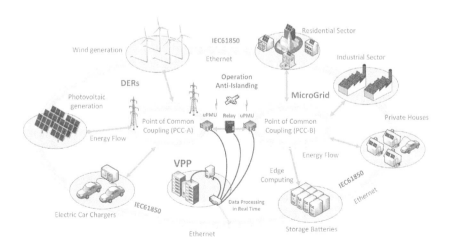

FIGURE 10.3 Control for microgrid interconnection.

198   Virtual Power Plant Solution for Future Smart Energy Communities

More advanced methods use PMU or synchrophasor measurement units [26], which must be connected both on the microgrid, with a relay connection, and on the main grid side, to measure and compare the phase angles of the two grids (Figure 10.3). These are devices capable of analysing voltage and current phasors synchronously with a global time reference. PMU measurements record grid conditions with high accuracy and provide stability information that can help to act as an Anti-Islanding system with the control of the Protection Relay. The idea is that in the event of a transient instability, the system can remain in Islanding mode until it is stabilised by a new connection to the utility grid.

Therefore, with these virtual synchronisation capabilities of frequency, voltage, and phase angle of the microgrid, it is possible to control these parameters in real time to improve operation and act as an Anti-Islanding avoiding a permanent disconnection from the grid. With a data bus working in real time, with advanced analysis capabilities and intelligent control of the inverters, it is possible to optimise the operation of the microgrids within the distribution network [27].

This method works when most of the power generated is AC, but in a microgrid with more than 40% of DC generators, new distributed control techniques with submillisecond synchronisation are needed [6]. At this level, it is important to have a synchronism standard such as TSN to extend the synchronism of PMUs to other equipment in charge of energy management in the various sectors.

On the other hand, for a VPP, the control of the electrical interconnection at the Points of Common Coupling (PCC-A and PPC-B) between the DERs and the microgrid takes on a fundamental role and has increased the complexity of the management of the electricity system, affecting the complexity of the VPPs [10, 28]. This is particularly true in distribution areas, where the passive model of energy consumption is giving way to bidirectional flows.

The new generation of systems for the management of VPPs is based on Embedded Systems located near to the generation systems to foster distributed processing and Edge Computing.

10.4   REAL-TIME PTP ARCHITECTURE FOR DISTRIBUTED GENERATION SYSTEM SYNCHRONISATION

The PTP or IEEE 1588 v2 standard was developed for the synchronisation of equipment in industrial areas. PTP is an evolution of the NTP protocol based on the Ethernet network, but with greater guarantees of stability. This protocol guarantees clock synchronisation errors of approximately 1 ms. For systems dedicated to distributed generation, this standard guarantees a proper correlation of all measurements in a power generation plant. Synchronisation with PTP is a substantial improvement for detecting disturbances and responding with accurate control in protection systems. This type of response is essential in other industrial and scientific sectors. There is a significant body of work that combines the PTP standard with other synchronisation techniques to increase accuracy. A very prominent one is based on the White Rabbit protocol [29] developed by CERN (European Council for Nuclear Research) researchers that combines Synchronous Ethernet SyncE and PTP to achieve subnanosecond accuracy and picoseconds precision of synchronisation.

Distributed Synchronism System Based on TSN and PTP

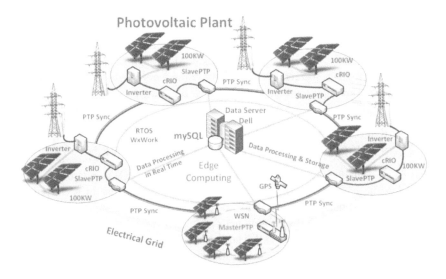

FIGURE 10.4 Synchronisation and capture the meteorological and electrical parameters with PTP.

In this section, an architecture for the synchronisation, capture, and processing of measurements with a distributed monitoring system is analysed. As an example, the design for a photovoltaic plant is shown in Figure 10.4. The PTP standard is used to achieve synchronisation of the internal clocks close to 50 μs [12]. Five devices are involved using the RTOS called VxWork, specially designed for embedded systems. With this RTOS, all the magnitudes equivalent to a power quality (PQ) meter are captured and processed simultaneously in the four cRIOs right at the transition of the second.

One of the cRIOs includes a GPS and acts as MasterPTP (bottom of Figure 10.4). The GPS module allows all wireless measurements to be time-stamped on a universal basis. In addition, this same cRIO acts as MasterPTP [30] to synchronise the clocks of the four cRIO systems installed in a 400 kW sector of the PV plant. The MasterPTP also acts as a wireless receiver for measurements of the DC levels generated by 12 PV trackers.

This type of architecture can be analysed in depth in [31]. The proposed system synchronises all equipment with PTP and the captured and processed parameters are stored on a server installed in the PV plant itself. This procedure ensures a rapid response to a critical type of disturbance. The server stores all data in MySQL. This is a relational database type not specially designed for time series storage, but it is the most widely used open-source database type. For remote access to this database, a RESTful service based on Java technology is used. This implementation allows universal access to the data as it can be transmitted and requested using the HTTP protocol, or in the exposed case HTTPS, to encrypt the communication. In addition, to restrict access only to authorised users, a password authentication layer was implemented. This is possible because the RESTful service provides flexible access to data and its ability to integrate with desktop applications.

This Edge-Database architecture is specially designed for Edge-Computing Topology. In energy management systems, distributed processing and local storage of time series allow working with event detection techniques with the capacity to anticipate critical situations. Latency is therefore reduced throughout the process, from the instant of capture to the instant of storage of the parameters in a database. This research work represents an important initiative with near real-time processing and storage.

10.5 ARCHITECTURE WITH MPMU AND TSN STANDARD AS KEY COMPONENTS IN THE FUTURE OF VPPs

IEEE bus models facilitate the study of power flows [32]. TSN standard allows synchronising the capture of critical events when using equipment in a conventional Ethernet network. In this context, the use of such a high-quality framework for VPPs to work simultaneously with distribution network models using real data obtained from the different facilities is justified [33]. The aim is to facilitate the diagnosis of faults and to analyse their propagation in the networks. Various research groups aim to maintain a data flow between the physical and synthetic installation in a deterministic and real-time manner. Using HIL technology [11] models can evolve with latencies in line with the real installation. This technology is gaining momentum for the development of digital twins [34]. It is under constant development in the energy sector [35] and is closely related to state-of-the-art technologies for the synchronisation of captures and events.

The TIC240 research group of the University of Córdoba proposes the extension of very high quality synchronism from μPMUs to other equipment. This allows to correlate the events produced within the generation installation with those that are being propagated in the distribution network. This proposal can be seen in a schematic way in Figure 10.5 and in [3]. To achieve this goal, a technological improvement of

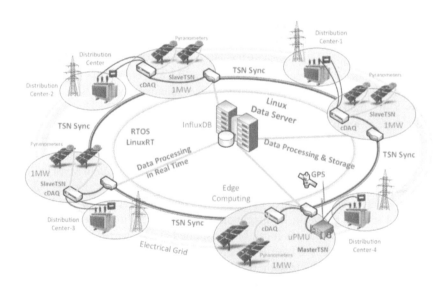

FIGURE 10.5 Monitoring system with TSN synchronism for VPP.

Distributed Synchronism System Based on TSN and PTP

μPMUs is developed with the integration of TSN-based synchronism [3] and real-time data transfer based on the IEC 61850 standard. It is an alternative to other types of proposals analysed by Agarwal et al. in [18], which present the design and implementation of an energy metering system with a DDS-based real-time communication platform with TSN synchronism. To transfer the three-phase measurements, the open-source DDS protocol widely developed by the company RTI [36] is leveraged by configuring the network data traffic according to specific quality of service (QoS) profiles, obtaining low packet loss and low latency with synchronisation and prioritisation of data in the network.

To corroborate this proposal, other publications that analyse the possibilities of μPMUs for advanced management of power distribution networks are discussed. For example, in 2018, Mohsenian-Rad et al. [37] published a study that analyses and justifies the use of PMUs in power distribution networks. Specifically, it focuses on the use of μPMUs, units that continuously measure time-referenced quantities as well as voltage and current phase angles of the overall system. With the use of μPMUs, real-time measurements with higher resolution and accuracy are achieved, and the diagnosis of possible faults in the network is facilitated by the higher level of visibility and monitoring of the distribution network. These units, together with the potential of Big Data, make it possible to remotely monitor hundreds of pieces of equipment in the grid, thus constituting a true awareness of distributed energy systems.

The tests conducted in the above-mentioned paper show that μPMU measurements provide much more detail on voltage fluctuations (e.g., momentary voltage drops of about 200 ms) compared to the same measurements made by standard Supervisory Control and Data Acquisition (SCADA) meters. In these cases, the key to identifying the root cause of voltage sags is the high temporal resolution as well as the time synchronisation of μPMU measurements. Since it allows differentiating between load- and grid-induced voltage sags or to determine more precisely those points in the generation and distribution network that are potentially susceptible to faults. The data collected can lead to the identification of anomalies that are not yet significant enough to cause a service interruption, but, if not addressed quickly enough, are likely to cause failures shortly.

The tests and data analysis carried out in the aforementioned article determine that μPMUs provide current and voltage measurements with higher resolution and accuracy to facilitate a level of monitoring in the distribution network that is currently difficult to achieve, allowing at the same time to detect future faults and, consequently, to establish a methodology for applying early predictive and corrective actions, preventing service interruptions to consumers [38].

With these expectations, μPMUs will contribute to maintaining a single, high-quality time frame [39, 40]. The μPMU will extend the synchronism to the rest of the measurement systems acting as MasterTSN. It is essential to evaluate the technical feasibility of this method according to the specifications of the IEEE C37.118.1a-2014 [41] standard that relaxes the requirements related to the specifications for phasor monitoring [42] included in IEEE C37.118.1-2011 [42]. The μPMU System Test and Calibration Guide for Power System Protection and Control [43] includes a methodology to verify the feasibility of a distributed synchronisation system. This guide is based on the IEEE Std C37.238™ 2011 standard [43], which defines the specifications of the IEEE 1588-2008 [30] for use in power systems. This guideline is appropriate for the proposal presented, taking into account that the IEEE 802.1AS-2011 [5] standard is based on the previous one.

10.6 THE IEC 61850 STANDARD AS A REFERENCE FRAMEWORK FOR VPPs

IEC 61850 was conceived as a standard for the substation environment, but later expanded its coverage to other systems involved in the management and control of power generation and transmission and included subsections such as IEC 61850-90-5 for the exchange of synchrophasor information in wide area network (WAN) environments, or IEC61850-90-12 describing WANs in the electric utility environment, focusing on parameters such as protocols, existing networks, transmission media, quality of service, redundancy, latency, and security.

For VPPs, WANs must interconnect the different microgrids that are bounded with local area networks (LANs), as in Figure 10.6, to facilitate the exchange of messages for control, management, and protection. At the local level, the different components of the microgrids exchange sample values (SVs) messages for electrical parameter measurements and GOOSEs messages for event detection. For the joint control and management of different microgrids, a message exchange similar to the previous one but adapted to the WAN environment is established through routed sampled values (R-SV) and R-GOOSE messages.

Reducing transfer times between two systems connected to the same distribution network requires a protocol that operates with a specially designed architecture. In the case of IEC 61850, it was designed to achieve very low latency messages, such as GOOSE-type messages, defined for asynchronous event transfer. The UDP/IP multicast protocol was chosen, because it allows much faster packet transfer and

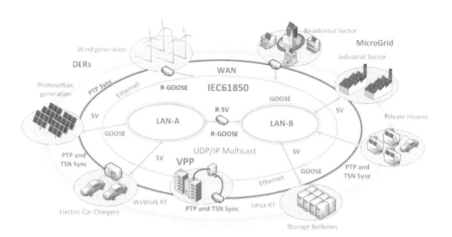

FIGURE 10.6 Communications with the IEC61850 standard for VPPs.

Distributed Synchronism System Based on TSN and PTP   **203**

the use of RTOS systems with the IEC 61850 standard is recommended depending on several factors such as data type and frame length. The paper [13] highlights some very relevant aspects of this standard such as the semantic definition for DERs and analyses the response times for processing new information nodes. This is a key factor because it mainly affects nodes with complex structures, such as those dedicated to complex event processing (CEP) and messages designed to achieve interoperability between systems that contribute to the overall stability of the network.

10.7 NEW ARCHITECTURE FOR CAPTURING, PROCESSING, AND MANAGING EDGE DATABASES

In the context of VPPs, there is a need for technologies that enable energy digitisation based on very high-performance synchronisation using quality standards. The management of energy generation and consumption in a given geographical area must be carried out through monitoring systems capable of detecting the origin and analysing the propagation of disturbances and of identifying the areas with supply that may be affected [44, 45]. The aim is to ensure accurate control of the bidirectional flow of energy, allowing various generation systems to cooperate to compensate for imbalances due to intermittent generation or fluctuating demand. These objectives can be achieved by using a type of technology that allows data to be captured, processed, and stored very efficiently and with a high degree of reliability. Edge-cloud computing techniques are commonly used for instantaneous detection of disturbances ensuring a rapid response to a potential power supply problem [46]. As mentioned above, PMUs, and more specifically μPMUs, allow analysis with a much higher temporal resolution [37]. They also allow the transmission of up to 120 phasors per second, but the management of such a large volume of data greatly complicates storage, management, and analysis with relational databases. The strategy currently being pursued is a two-pronged approach. On the one hand, with mass storage in edge-database and, on the other hand, with the use of a database manager specially designed for the optimisation of time series data. InfluxDB is an open-source tool that was developed for this purpose. To interpret the data coming from the μPMUs (Figure 10.6), an intermediary is needed. In this discipline, they are called agents, and the most commonly used for influxDB is Telegraf, although for phasor meters the most suitable is OpenPDC, developed by the Grid Protection Alliance (GPA) [47].

The method followed with an architecture as shown in Figure 10.7 is to process the data very close to the source and to have a local storage system to facilitate immediate analysis "cloudlets computing". Such a storage system should have a high-resolution data buffer but with limited size. The proposed architecture also envisages remote analysis with cloud computing with a mirror database but with less temporal resolution. It is common to use the FIWARE platform for centralised management of data from several distributed databases. This method can contribute to the coordinated management of small "power plants", so that they cooperate acting as a VPP.

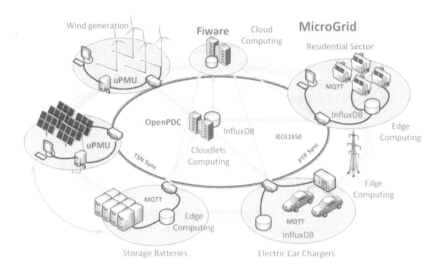

FIGURE 10.7 Edge-computing strategy for VPPs.

REFERENCES

1. M. Chen and H. V. Poor, "High-Frequency Power Electronics at the Grid Edge: A Bottom-Up Approach toward the Smart Grid," *IEEE Electrif. Mag.*, vol. 8, no. 3, pp. 6–17, Sep. 2020.
2. K. B. Stanton, "Distributing Deterministic, Accurate Time for Tightly Coordinated Network and Software Applications: IEEE 802.1AS, the TSN Profile of PTP," *IEEE Commun. Stand. Mag.*, vol. 2, no. 2, pp. 34–40, Jun. 2018.
3. V. Pallarés-López, R. J. Real-Calvo, S. del R. Jiménez, M. González-Redondo, I. Moreno-García, and I. Santiago, "Monitoring of Energy Data with Seamless Temporal Accuracy Based on the Time-Sensitive Networking Standard and Enhanced μPMUs," *Appl. Sci.* vol. 11, no. 19, p. 9126, Sep. 2021.
4. "IEEE Draft Standard Profile for Use of IEEE Std. 1588 Precision Time Protocol in Power System Applications." pp. 1–67, 2011.
5. S. IEEE 802.1AS, "IEEE Std 802.1AS-2011 – Time and Synchronization for Time-Sensitive Applications in Bridged Local Area Networks," in *IEEE Standards*, no. March 30, 2011, pp. 1–147.
6. M. Asawa, B. Murphy, S. Bose, "Whitepaper: Grid-Synchronized, 100% renewable, Business-Ready, Digital Microgrid" [Online]. Available: https://medium.com/@wiprodigital/whitepaper-grid-synchronized-100-renewable-business-ready-digital-microgrid-by-dr-5ecf5e655f30. [Accessed: 09-Jul-2020].
7. T. Popławski, S. Dudzik, P. Szeląg, and J. Baran, "A Case Study of a Virtual Power Plant (VPP) as a Data Acquisition Tool for PV Energy Forecasting," *Energies* vol. 14, no. 19, p. 6200, Sep. 2021.
8. B. Behi, A. Baniasadi, A. Arefi, A. Gorjy, P. Jennings, and A. Pivrikas, "Cost–Benefit Analysis of a Virtual Power Plant Including Solar PV, Flow Battery, Heat Pump, and Demand Management: A Western Australian Case Study," *Energies* vol. 13, no. 10, p. 2614, May 2020.
9. P. Pal et al., "Optimal Dispatch Strategy of Virtual Power Plant for Day-Ahead Market Framework," *Appl. Sci.* vol. 11, no. 9, p. 3814, Apr. 2021.

10. E. A. Bhuiyan, M. Z. Hossain, S. M. Muyeen, S. R. Fahim, S. K. Sarker, and S. K. Das, "Towards Next Generation Virtual Power Plant: Technology Review and Frameworks," *Renew. Sustain. Energy Rev.*, vol. 150, p. 111358, Oct. 2021.
11. S. Essakiappan et al., "A Multi-Site Networked Hardware-in-the-Loop Platform for Evaluation of Interoperability and Distributed Intelligence at Grid-Edge," *IEEE Open Access J. Power Energy*, vol. 8, pp. 460–471, Aug. 2021.
12. A. Moreno-Munoz, V. Pallares-Lopez, J. J. Gonzalez de la Rosa, R. Real-Calvo, M. Gonzalez-Redondo, and I. M. Moreno-Garcia, "Embedding Synchronized Measurement Technology for Smart Grid Development," *IEEE Trans. Ind. Informatics*, vol. 9, no. 1, pp. 52–61, Feb. 2013.
13. M. J. Gonzalez-Redondo, A. Moreno-Munoz, V. Pallares-Lopez, and R. J. Real-Calvo, "Influence of Data-Related Factors on the Use of IEC 61850 for Power Utility Automation," *Electr. Power Syst. Res.*, vol. 133, pp. 269–280, 2016.
14. "IEEE 1815-2010 – IEEE Standard for Electric Power Systems Communications – Distributed Network Protocol (DNP3)." [Online]. Available: https://standards.ieee.org/standard/1815-2010.html. [Accessed: 05-Nov-2021].
15. V. Pallares-Lopez, A. Moreno-Muñoz, M. Gonzalez-Redondo, R. Real-Calvo, I. M. García, and J. J. G. De La Rosa, "Synchrophasor Integration in IEC 61850 Standard for SmartGrid and Synchronism with PTP-Base System," in *Proceedings of the 2011 6th IEEE Conference on Industrial Electronics and Applications, ICIEA 2011*, 2011, pp. 1507–1512.
16. "Digital Twins for Power Systems | Hardware-in-the-Loop | OPAL-RT." [Online]. Available: https://www.opal-rt.com/digital-twins/. [Accessed: 05-Nov-2021].
17. J. Wang, J. Simpson, R. Yang, B. Palmintier, S. Tiwari, and Y. Zhang, "Hardware-in-the-Loop Evaluation of an Advanced Distributed Energy Resource Management Algorithm," *2021 IEEE Power Energy Soc. Innov. Smart Grid Technol. Conf. ISGT 2021*, Feb. 2021.
18. T. Agarwal, P. Niknejad, M. R. Barzegaran, and L. Vanfretti, "Multi-Level Time-Sensitive Networking (TSN) Using the Data Distribution Services (DDS) for Synchronized Three-Phase Measurement Data Transfer," *IEEE Access*, vol. 7, pp. 131407–131417, 2019.
19. "Products – Connext DDS: Connectivity Framework | RTI." [Online]. Available: https://www.rti.com/products. [Accessed: 28-Apr-2020].
20. DDS Foundation, "What is DDS?"[Online]. Available: https:// www.dds-foundation.org/what-is-dds-3/. [Accessed: 28-Apr-2020].
21. M. Smith, "Evaluation of IEEE 802.1 Time Sensitive Networking Performance for Microgrid and Smart Grid Power System Applications," *Masters Theses*, Aug. 2018.
22. "The Industrial Internet Consortium Announces Microgrid Testbed Results and Publishes Whitepaper - Wipro." [Online]. Available: https://www.wipro.com/digital/news/the-industrial-internet-consortium-announces-microgrid-testbed-results-and-publishes-whitepaper/. [Accessed: 28-Apr-2020].
23. "Industrial Internet Consortium." [Online]. Available: https://www.iiconsortium.org/. [Accessed: 28-Apr-2020].
24. R. Real-Calvo, A. Moreno-Munoz, J. Gonzalez-De-La-Rosa, V. Pallares-Lopez, M. Gonzalez-Redondo, and I. Moreno-Garcia, "An Embedded System in Smart Inverters for Power Quality and Safety Functionality," *Energies*, vol. 9, no. 3, p. 219, Mar. 2016.
25. A. E. Saldaña-González, A. Sumper, M. Aragüés-Peñalba, and M. Smolnikar, "Advanced Distribution Measurement Technologies and Data Applications for Smart Grids: A Review," *Energies* vol. 13, no. 14, p. 3730, Jul. 2020.
26. "Applications Synchrophasor Technology: Program Impacts: Recovery Act | SmartGrid. gov." [Online]. Available: https://www.smartgrid.gov/recovery_act/program_impacts/applications_synchrophasor_technology.html. [Accessed: 28-Apr-2020].

27. R. J. Real-Calvo, A. Moreno-Munoz, V. Pallares-Lopez, M. J. Gonzalez-Redondo, and J. M. Flores-Arias, "Intelligent Electronic Device for the Control of Distributed Generation," in *IEEE International Conference on Consumer Electronics – Berlin, ICCE-Berlin*, 2015.

28. K. Mahmud, B. Khan, J. Ravishankar, A. Ahmadi, and P. Siano, "An Internet of Energy Framework with Distributed Energy Resources, Prosumers and Small-Scale Virtual Power Plants: An Overview," *Renew. Sustain. Energy Rev.*, vol. 127, p. 109840, Jul. 2020.

29. E. L. English *et al.*, "The WRITE (White Rabbit for Industrial Timing Enhancement) Project Update," in *Proceedings of the 52nd Annual Precise Time and Time Interval Systems and Applications Meeting*, pp. 146–166, Jan. 2021.

30. "1588-2008 – IEEE Standard for a Precision Clock Synchronization Protocol for Networked Measurement and Control Systems – Redline – IEEE Standard." [Online]. Available: https://ieeexplore.ieee.org/document/7949184. [Accessed: 10-Jan-2020].

31. I. M. Moreno-Garcia *et al.*, "Real-Time Monitoring System for a Utility-Scale Photovoltaic Power Plant," *Sensors* vol. 16, no. 6, p. 770, May 2016.

32. S. Tiwari, M. A. Ansari, K. Kumar, S. Chaturvedi, M. Singh, and S. Kumar, "Load Flow Analysis of IEEE 14 Bus System Using ANN Technique," in *2018 International Conference on Sustainable Energy, Electronics and coMputing System, SEEMS 2018*, 2019.

33. A. Rasheed, O. San, and T. Kvamsdal, "Digital Twin: Values, Challenges and Enablers from a Modeling Perspective," *IEEE Access*, vol. 8, pp. 21980–22012, 2020.

34. P. Jain, J. Poon, J. P. Singh, C. Spanos, S. R. Sanders, and S. K. Panda, "A Digital Twin Approach for Fault Diagnosis in Distributed Photovoltaic Systems," *IEEE Trans. Power Electron.*, vol. 35, no. 1, pp. 940–956, Jan. 2020.

35. H. Pan, Z. Dou, Y. Cai, W. Li, X. Lei, and D. Han, "Digital Twin and Its Application in Power System," in *2020 5th International Conference on Power and Renewable Energy, ICPRE 2020*, 2020, pp. 21–26.

36. "Data Distribution Service (DDS) for Complex Systems | RTI." [Online]. Available: https://www.rti.com/products/dds-standard. [Accessed: 07-Nov-2021].

37. H. Mohsenian-Rad, E. Stewart, and E. Cortez, "Distribution Synchrophasors: Pairing Big Data with Analytics to Create Actionable Information," *IEEE Power Energy Mag.*, vol. 16, no. 3, pp. 26–34, May 2018.

38. M. Adamiak, W. Premerlani, and B. Kasztenny, "Synchrophasors: Definition, Measurement, and Application."

39. R. Ghiga, Q. Wu, K. Martin, W. El-Khatib, L. Cheng, and A. H. Nielsen, "Steady-State PMU Compliance Test Under C37.118.1a™-2014," in *IEEE PES Innovative Smart Grid Technologies Conference Europe*, 2016, pp. 1–6.

40. R. Ghiga, Q. Wu, K. Martin, W. Z. El-Khatib, L. Cheng, and A. H. Nielsen, "Dynamic PMU compliance test under C37.118.1a™-2014," in IEEE Power and Energy Society General Meeting, 2015.

41. "IEEE Standard for Synchrophasor Measurements for Power Systems – Amendment 1: Modification of Selected Performance Requirements," *IEEE Std C37.118.1a-2014 (Amendment to IEEE Std C37.118.1-2011)*, 2014.

42. K. E. Martin, "Synchrophasor Measurements Under the IEEE Standard C37.118.1-2011 With Amendment C37.118.1a," *IEEE Trans. Power Deliv.*, vol. 30, no. 3, pp. 1514–1522, Jun. 2015.

43. "IEEE Guide for Synchronization, Calibration, Testing, and Installation of Phasor Measurement Units (PMUs) for Power System Protection and Control IEEE Power and Energy Society," C37.242-2013, 2013.

44. A. Tahabilder, P. K. Ghosh, S. Chatterjee, and N. Rahman, "Distribution System Monitoring by using Micro-PMU in Graph-Theoretic Way," in *2017 4th International Conference on Advances in Electrical Engineering (ICAEE)*, 2017, pp. 159–163.
45. M. Jamei *et al.*, "Anomaly Detection using Optimally Placed µPMU Sensors in Distribution Grids," *IEEE Trans. Power Syst.*, vol. 33, pp. 3611-3623, 2018.
46. D. Sodin, U. Rudež, M. Mihelin, M. Smolnikar, and A. Čampa, "Advanced Edge-Cloud Computing Framework for Automated PMU-Based Fault Localization in Distribution Networks," *Appl. Sci.*, vol. 11, no. 7, p. 3100, Mar. 2021.
47. "Grid Protection Alliance – Synchrophasor Products." [Online]. Available: https://www.gridprotectionalliance.org/productsPhasor.asp#PDC. [Accessed: 08-Nov-2021].

11 Complementarity and Flexibility Indexes of an Interoperable VPP

Habib Nasser and Dah Diarra
RDIUP, Les Mureaux, France

CONTENTS

11.1 Introduction 210
11.2 Complementarity 210
    11.2.1 Overview 210
    11.2.2 Proposed Method 212
11.3 Flexibility 213
    11.3.1 Overview 213
    11.3.2 Proposed Method 214
11.4 Case Study 215
    11.4.1 Complementarity 216
    11.4.2 Flexibility 217
11.5 Conclusion and Future Work 218
Acknowledgment 219
References 219

NOMENCLATURE

API	application programming interface
CI	complementarity index
CL	controllable loads
DER	distributed energy resource
ES	energy source
FI	flexibility index
RES	renewable energy source
S	energy storage
VPP	virtual power plant
VRE	variable renewable energy

DOI: 10.1201/9781003257202-12

209

11.1 INTRODUCTION

The field of energy systems has received special attention in this decade from the scientific community around the world. This is due to the urgent need for a sustainable transition to a fossil-free energy system with low greenhouse gas emission which will ensure a safe environment. Renewable energy systems are the main component of this energy source (ES) transition. So most proposed studies for the transition are based on how to integrate renewable ESs (RESs) to existing grid systems and then reduce the use of fossil ESs. As a result, during this decade, RES has gained an increasing penetration over fossil ESs [1]. But wind and solar have a high share of RES integration [1] with both being VRE (variable renewable energy) generators, and this results in a huge challenge in planning, transmission and distribution of energy. It then becomes more obvious to think about a better way of analyzing the entire grid composition and its behaviors. Analyzing the flexibility of the grid and the complementarity of virtual power plant (VPP) components can provide a better understanding of its behaviors from intraday to years.

In our study and according to [2], a VPP is a practical concept that combines various distributed energy resources (DERs) to improve energy management efficiency and facilitate energy trading. The main purposes of a VPP are to enhance and optimize power generation, as well as trading or selling power on the electricity market. A VPP can be considered to be performant in one hand if its components form together an optimal combination based on a defined criteria, and on the other hand, it has the ability to handle the variability or fluctuation and uncertainty on both generation and consumption side. This chapter will propose two metrics called complementarity index (CI) and flexibility index (FI) to determine, respectively, the level of complementarity and flexibility of a given standalone and interoperable VPP.

As VPP can be composed of RES and non-RES energy generators, this study will focus on VPP composed of only RES as energy generators, storage systems and controllable loads (CLs).

We also assume that the VPP is installed in a standalone manner, which allows it to fully control the system and provide better management in transmission and distribution. These kinds of VPP are useful for island cities.

In the next sections, we will first present an overview of CI in the literature, then the proposed metric for CI calculation. After that, similar study will be carried out for the FI and then will present a case study in order to experiment with these indexes and assess preliminary results, and finally, future work and conclusion are described.

11.2 COMPLEMENTARITY

11.2.1 Overview

The choice of VPP energy resources and other components is a critical investment decision for managers. This task needs proper information that states the context and gives better understanding of the current need of the VPP. The evaluation of VPP

Complementarity and Flexibility Indexes of an Interoperable VPP **211**

resources complementarity between each other and especially the complementarity between ESs allow one to make an appropriate decision. The CI is a popular evaluation metric that determines how much complementary VPP resources or components provide.

According to [3], complementarity should be considered as the capability of energy components of working in a complementary way, and it can be observed in time, space and jointly in both domains. ES complementarity can be mainly divided into spatial, temporal and spatiotemporal complementarity. One or more ESs are considered to be spatially complementary when their production complements each other in different space or regions. And two or more ESs are considered to be temporarily complementary when their production complements each other at different periods of time in the same region or space. Spatiotemporal complementarity consists of using both temporal and space complementarity during the analysis. Beluco et al. in [4] defined complementarity in the energy field as the capacity of two or more ESs to be complementary available between them. And then propose a CI K defined by equation (11.1), where K_t, K_e and K_a are partial indexes measuring complementary based on time, energy and amplitude, respectively.

$$K = K_t \times K_e \times K_a \tag{11.1}$$

Time-complementarity K_t is defined by equation (11.2), where D and d are, respectively, the number of maximum and minimum days of availability of a given ES.

$$K_t = \frac{|d_1 - d_2|}{\sqrt{|D_1 - d_1||D_2 - d_2|}} \tag{11.2}$$

Energy-complementarity K_e is defined by equation (11.3) where E represents the total energy produced by a given source during the year.

$$K_e = 1 - \frac{|E_1 - E_2|}{|E_1 + E_2|} \tag{11.3}$$

Amplitude-complementarity K_a is defined by equation (11.4), where δ_1 and δ_2 represent a score resulting from equation (11.5), where E_{max}, E_{min} and E_{av} represent, respectively, in equation (11.5) maximum, minimum and average value of energy availability for a given source.

$$K_a = \begin{cases} 1 - \dfrac{(\delta_1 - \delta_2)^2}{(1 - \delta_2)^2}, & if \quad \delta_1 \le \delta_2 \\[4mm] \dfrac{(1 - \delta_2)^2}{(1 - \delta_2)^2 + (\delta_1 - \delta_2)^2}, & if \quad \delta_1 \ge \delta_2 \end{cases} \tag{11.4}$$

$$\delta = 1 + \frac{E_{max} - E_{min}}{E_{av}} \tag{11.5}$$

212   Virtual Power Plant Solution for Future Smart Energy Communities

Beluco et al. used CI in [4] to develop a complementarity map of wind and solar energy based on Rio Grande do Sul energy data. Also in [5], it was used to evaluate complementarity between wind, hydropower and solar energy based on Rio Grande do Sul energy data. In [6], maps of correlation between wind and water were used to evaluate the complementarity between both ESs based on Brazilian territory energy data.

In this chapter, we will propose a novel evaluation metric to calculate the CI of a VPP that will take into account ES, available storage, CL capacity and consumption loads. The methodology presented is based on the assumption that the VPP is or will works in a standalone environment where it handles all the processes from energy generation to consumption without any outside influence.

11.2.2 PROPOSED METHOD

Given a VPP, V, with a set of ESs, $ES = es_1, es_2,..., es_n$, a set of energy storage, $S = s_1, s_2,..., s_n$ and a set of CL, $CL = cl_1, cl_2,..., cl_n$. V is defined as complementary VPP if for each element es_i of ES, there is another element es_j in ES that can compensate the energy loss or excess of es_i while responding to the VPP fluctuation or variation in both generation and consumption, and also if it's CL and storages can be used to regulate the network when total generation and total consumption do not match in a single point of time. CI allows us to determine this degree or level of complementarity.

This index is divided into three different parts: journey CI between ESs, seasonal CI between ESs and the index of tolerance of the VPP that represents the usability of the storage systems and CL to mitigate the gaps between demand-response. So the complementarity index, CI, is defined by equation (11.6), where $Card (ES)$ is a number of active ESs of the VPP.

$$CI = \begin{cases} \dfrac{1}{3}\left(CI_{journey} + CI_{seasonal} + CI_{tolerance}\right) \\ 0 \quad if \quad Card(ES) \leq 1 \end{cases} \tag{11.6}$$

The journey complementarity index, $CI_{journey}$, is defined by equation (11.7), where we iteratively look for highest negative correlation between es_h that represents the hourly energy production of a single ES of the VPP and the u_h which characterizes the hourly production of all ESs along a full day.

$$CI_{journey} = \sum_{es_h \in ES} \frac{min\left(0, argmin\left(corr\left(es_h, u_h\right)\right)\right)}{Card(ES)} \quad \forall u_h \in ES \tag{11.7}$$

Complementarity and Flexibility Indexes of an Interoperable VPP 213

The seasonal complementarity index, $CI_{seasonal}$, is defined by equation (11.8), where es_d and u_d are similar to es_h and u_h in $CI_{journey}$ but with daily data along the full year.

$$CI_{seasonal} = \sum_{es_d \in ES} \frac{min\left(0, argmin\left(corr(es_d, u_d)\right)\right)}{Card(ES)} \quad \forall u_d \in ES \tag{11.8}$$

The tolerance complementarity index, $CI_{tolerance}$, is defined by equation (11.9), where $load_t$ and $prod_t$ represent, respectively, the total energy production and consumption at a given time, N defines the number of observations, *storage* and *cl* represent, respectively, the total number of storage and CL capacity. ϵ_1 and ϵ_2 represent, respectively, the degree of consideration of the capacity of storage and CL over their total values. The $CI_{tolerance}$ is very important to distinguish the complementarity between VPPs with and without storage systems. Also, the capacity of storage will have an impact on the CI index of any VPP. The higher the capacity, the more complementarity weight will increase.

$$CI_{tolerance} = \sum_{t=1}^{N} \frac{1 - min\left(1, \dfrac{|load_t - prod_t|}{storage \times \epsilon_1 + cl \times \epsilon_2}\right)}{N} \tag{11.9}$$

11.3 FLEXIBILITY

11.3.1 OVERVIEW

With the increasing penetration of RESs (especially wind and solar) in power systems, the planning and control of the power system are becoming a huge challenge due to their uncertainty and variability. The capacity of the VPP to handle these uncertainty and variation during time is called flexibility. In [7], the term flexibility was defined as the ability of a power system to cope with variability and uncertainty in both generation and demand, while maintaining a satisfactory level of reliability at a reasonable cost over different time horizons. Another definition of the term in [8] was the ability of a power system to maintain continuous service in face of rapid and large swings in supply and demand.

FI is a metric to determine the level of flexibility of a power system. It can be used by VPP managers to monitor and improve the power system. Different metrics were proposed in the literature for its calculation. Berahmandpour, Montaser Kouhsari and Rastegar introduce in [9] a new FI for real-time operation incorporating wind farms. This index is based on up and down generation constraints and ramp rate limitations of each unit of the power system. The insufficient ramping resource expectation (IRRE) metric was proposed to measure power system flexibility for use in long-term planning [10]. Zhao, Zheng and Litvinov introduce a Boolean-based index indicating whether or not a power system's largest variation range is within a given target range [11].

214 Virtual Power Plant Solution for Future Smart Energy Communities

In this chapter, we propose an FI that takes into account VPP historical or forecasted ES production, available free and full storage capacity, ramping capacity and loads data in a given time horizon. A detailed explanation of the methodology will be presented in the next section.

11.3.2 PROPOSED METHOD

Consider a given VPP, V, composed of a $ES = es_1, es_2, \ldots es_n$ and $S = s_1, s_2, \ldots s_n$, which are, respectively, a set of energy generation sources that can be dispatchable (provide ramping capabilities) or intermittent and energy storage. V is said to be flexible at a given time t if the difference between generation and loads can be covered by the storage capacity or by ramping up or down some of its energy generation source even with some unexpected variations from loads or generations (mainly intermittent sources). The proposed FI is based on seeking for V deficiency to provide flexibility over time. The final FI of V in a specific interval is the mean FI of V at each time and defined by equation (11.10).

$$FI = \frac{1}{N} \sum_{t=1}^{N} Flex(t) \tag{11.10}$$

where $Flex(t)$ is the FI at time t. It is defined by equation (11.11). In order to quantify the level of the flexibility at a given time, consider a $E = \{(i_1, j_1), (i_2, j_2), \ldots, (i_n, j_n)\}$ a set of couples of random variations from loads and generations, $Flex(t)$ given by equation (11.11) is the mean of the FI at time t with every couple of variations. This index shows how well V handles uncertainty and variability from both generation and loads.

$$Flex(t) = \frac{1}{Card(E)} \sum_{(i,j) \in E} Flex(t, i, j) \tag{11.11}$$

where $Flex(t, i, j)$ determines the FI at time t according to i and j that represent, respectively, the variation generated from productions and loads. This index has a float basis, so its value can only be between 1 if flexible and 0 otherwise. $Flex(t, i, j)$ is then defined by equation (11.12) where we assume that the ramping up and down are superior to zero and the difference between production and consumption is never equal to zero.

$$Flex(t, i, j) = \begin{cases} min\left(1, \dfrac{full\_storage_t + ramp\_up_t}{|prod_t * i - load_t * j|}\right) & if \quad prod_t * i - load_t * j < 0 \\[4mm] min\left(1, \dfrac{free\_storage_t + ramp\_down_t}{|prod_t * i - load_t * j|}\right) & if \quad prod_t * i - load_t * j > 0 \end{cases} \tag{11.12}$$

where $full\_storage_t$ and $free\_storage_t$ determine, respectively, the total stored energy that can be consumed the available storage capacity at time t, $ramp\_up_t$ and $ramp\_down_t$

represent, respectively, the total ramping up and down capacity at time t, and $prod_t$ and $load_t$ are, respectively, total energy production and consumption at time t.

11.4 CASE STUDY

To test and assess these factors, an interoperable API was developed to facilitate the future demonstration in real-settings conditions. The API will have many modules (see Figure 11.1) in order to provide a tailored solution based on the requirements. One of these modules is responsible for data transfer or ingestion between the API and other external data providers. This module supports three main data transfer protocols such as Http(s), WebSocket and MQTT. Http protocol is used by API clients to send or retrieve data from the module where a new connection is established for each request. WebSocket protocol is used by clients to establish a connection with the module one time and then can send or receive data in real time without establishing new connection. The MQTT protocol is used to communicate with a shared MQTT server that does not belong to the API, but our module uses this server to listen for new data and also broadcast data to clients through this protocol. By providing these multiple communication protocols, any VPP including this API will be interoperable with different modules and paradigms.

In this case study, we use the "Île-de-France" region energy generation in 2020 dataset [12] to provide a real-world usage of the proposed CI and FI. For the sake of simplicity, only wind, solar and bioenergy are used in this study as clean energy

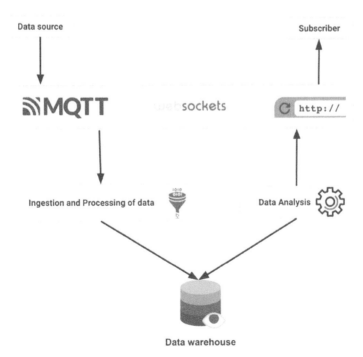

FIGURE 11.1 Interoperable cloud-based application.

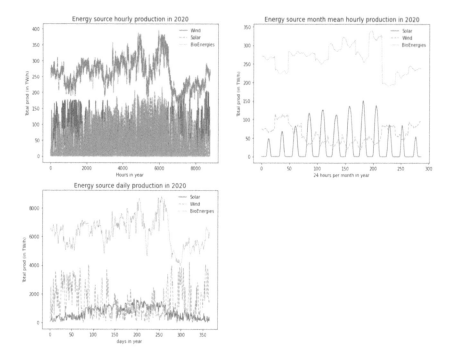

FIGURE 11.2 Energy source generations in 2020.

generation sources. Bioenergy is dispatchable and provides a ramping capacity that can be used to regulate the energy network. We also fine tune the load consumption according to the used ES as we did not use all ESs.

Figure 11.2 depicts different ways of visualizing each ES production capacity during the year of 2020, where the first column shows the total energy produced by each source per hour. We can see that a high share of the production is ensured by the bioenergy source. The second column shows the mean hourly total production of a given month of each source. We can notice the expected behaviors of the solar source which is only available at some hours during a day. Then the third column shows the total daily energy production per source, also wind sources decrease in spring and summer, whereas solar increase which can be considered as a kind of potential complementarity.

Figure 11.3 depicts the total energy generation resulting from aggregation of three sources and total energy consumption (loads). The first column displays, respectively, total generation and consumption per hours, whereas the second column the per day. We can notice that variations of consumption are more important than the production fluctuation.

11.4.1 COMPLEMENTARITY

In order to evaluate the level of complementarity of this power system in "Île-de-France", we computed the CI with different values of storage CL capacity. Table 11.1 describes the obtained results.

Complementarity and Flexibility Indexes of an Interoperable VPP 217

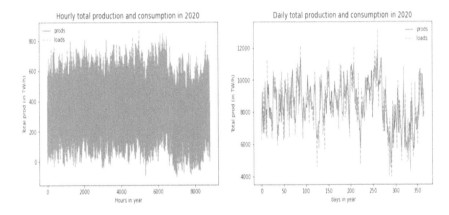

FIGURE 11.3 Energy total production and consumption in 2020.

As shown in Table 11.1, we can see that the couple (wind-solar) is more complementary than the couple (bioenergy-wind) which confirms that the VPP components have real impact on the complementarity. Moreover, we note that the power system provides a highly seasonal complementarity than journey, which can be seen in Figure 11.2. Also, the CI decreases with low-level storage and CL capacities decreases. One thing to remember is that the storage and CL are crucial to this power system, because if wind and solar variations exceed the ramping capacity, the entire system will go down.

11.4.2 FLEXIBILITY

To evaluate the flexibility of the system, we introduced various values of available storage capacities (0, 50, 100 MWh). This can be used as backup when the ramping capacity of the bioenergy is not able to balance the network. The normal production capacity of the installed bioenergy is 120 MW/h with a ramping up capacity of 60%

TABLE 11.1
Complementarity Index with Different Storage and CL Capacity

	ESs	Storage	Controllable loads	Journey CI	Season CI	Tolerance CI	CI
0	Wind, Solar	500	200	0.223689	0.445353	0.784346	0.484463
1	Wind, Solar	500	10	0.223690	0.445354	0.704162	0.457735
2	Wind, Solar	100	250	0.223691	0.445355	0.569307	0.412783
3	Wind, Solar	0	0	0.223692	0.445356	0.002277	0.223773
4	BioEnergy, Wind	500	200	0.209926	0.266876	0.786136	0.420979
5	BioEnergy, Wind	500	10	0.209927	0.266877	0.706618	0.394473
6	BioEnergy, Wind	100	250	0.209928	0.266878	0.572882	0.349895
7	BioEnergy, Wind	0	0	0.209929	0.266879	0.002049	0.159617

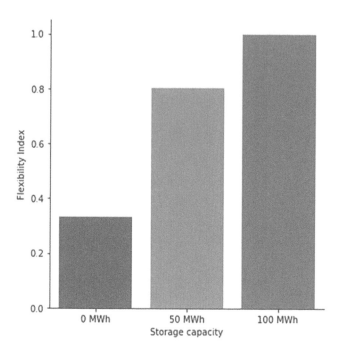

FIGURE 11.4 Flexibility index with different level of storage.

and ramping down capacity of 40%. Figure 11.4 shows the obtained result of the FI with different storage capacities.

As presented in Figure 11.4, we can notice that the actual ramping capacity (where the storage capacity is null and with only the ramping up [30% of 120 MWh] or down [15% of 120 MWh] capacities) did not provide enough flexibility to balance the power system when there are unexpected variations in energy production or consumption, however when the capacity of the storage increased, the FI rises significantly. Therefore, a VPP with a high share of variable renewable energy sources and low ramping capacities will require a high capacity of storage system to become flexible.

11.5 CONCLUSION AND FUTURE WORK

Recent years have been marked by an increasing penetration of renewable energy systems in power systems due to the desire to move from fossil ES to fossil-free ES as soon as possible. The integration of RESs comes with planning and control challenges, as the high share of renewable sources actually integrated are wind and solar that are both variable and nondispatchable. Defining key metrics that help power system (VPP) managers to make proper planning and control very accurately will allow RES to be easily integrated among existing grids. In this chapter, we propose methodology beyond the state of the art to calculate the level of complementarity (CI) and flexibility indexes (FI) of a power system. The preliminary results show that the combination of energy sources impact the complementarity of VPP. Regarding

Complementarity and Flexibility Indexes of an Interoperable VPP   219

the flexibility index, the ramping and especially the storage systems are crucial to improve the flexibility of a green power system.

CI calculates how complementary are the power system components such as ES, storage capacity and CL capacity. The methodology defines two ESs as complementary if they have a negative correlation over time. FI determines how well the power system handles unexpected behaviors from both generation and consumption side. This index is calculated by determining a set of variabilities in generation and consumption over time and then seeking where the power system failed.

The provided FI and CI do not consider the price fluctuation and the configuration of the different used RESs in their equations. In future research, we intend to adapt the equation to one that use these parameters and also integrate fossil ESs in order to make tools to handle this transition phase.

ACKNOWLEDGMENT

This work was carried out by the H2020 VPP4ISLANDS Project co-funded by the European Commission under the agreement GA No. 957852.

REFERENCES

1. Semich Impram, Secil V. Nese, Bülent Oral, "Challenges of renewable energy penetration on power system flexibility: A survey," Energy Strategy Reviews, Volume 31, 2020, 100539, ISSN 2211-467X, https://doi.org/10.1016/j.esr.2020.100539.
2. Hossein M. Rouzbahani, Hadis Karimipour, Lei Lei, "A review on virtual power plant for energy management," Sustainable Energy Technologies and Assessments, Volume 47, 2021, 101370, ISSN 2213-1388, https://doi.org/10.1016/j.seta.2021.101370.
3. J. Jurasz, F.A. Canales, A. Kies, M. Guezgouz, A. Beluco. 2020. "A review on the complementarity of renewable energy sources: Concept, metrics, application and future research directions." https://www.sciencedirect.com/science/article/pii/S0038092X19311831.
4. Alexandre Beluco, Paulo Kroeff de Souza, Arno Krenzinger. 2008. "A dimensionless index evaluating the time complementarity between solar and hydraulic energies." https://www.sciencedirect.com/science/article/pii/S0960148108000116.
5. Marcos Bagatini, Mariana G. Benevit, Alexandre Beluco, Alfonso Risso. 2017. "Complementarity in time between hydro, wind and solar energy resources in the State of Rio Grande do Sul, in Southern Brazil." https://www.scirp.org/journal/paperinformation.aspx?paperid=78679.
6. Mauricio P. Cantão, Marcelo R. Bessa, Renê Bettega, Daniel H.M. Detzel, João M. Lima. 2017. "Evaluation of hydro-wind complementarity in the Brazilian territory by means of correlation maps." https://doi.org/10.1016/j.renene.2016.10.012.
7. Danish and European Experiences. 2015. "Flexibility in the power system." https://ens.dk/sites/ens.dk/files/Globalcooperation/flexibility_in_the_power_system_v23-lri.pdf.
8. Georgios Papaefthymiou, Katharina Grave, Ken Dragoon. 2014. "Flexibility options in electricity systems." https://www.ourenergypolicy.org/wp-content/uploads/2014/06/Ecofys.pdf.
9. H. Berahmandpour, S. Montaser Kouhsari, H. Rastegar. 2019. "A new flexibility index in real time operation incorporating wind farms." https://www.researchgate.net/publication/334994586_A_New_Flexibility_Index_in_Real_Time_Operation_Incorporating_Wind_Farms.

10. E. Lannoye, D. Flynn, M. O'Malley. 2012. "Evaluation of power system flexibility." In IEEE Transactions on Power Systems. https://ieeexplore.ieee.org/document/6125228.
11. J. Zhao, T. Zheng, E. Litvinov. 2016. "A unified framework for defining and measuring flexibility in power systems." https://ieeexplore.ieee.org/document/7024953.
12. France energy data by RTE. https://www.rte-france.com/eco2mix.

12 Experience of Building a Virtual Power Plant of the Island of El Hierro

Oleksandr Novykh[1], Juan Albino Méndez Pérez[2], Benjamín González-Díaz[1], Jose Francisco Gomez Gonzalez[1], Igor Sviridenko[3], and Dmitry Vilenchik[4]

[1]Departamento de Ingeniería Industrial, Universidad de La Laguna, Avda. Astrofísico Francisco Sánchez, La Laguna, Tenerife, Spain
[2]Departamento de Ingeniería Informática y de Sistemas, Universidad de La Laguna, Avda. Astrofísico Francisco Sánchez, La Laguna, Tenerife, Spain
[3]Department of Energy Facilities of Ships and Marine Structures, Maritime Institute, State University of Sevastopol, Sevastopol, Russia
[4]Frish Technologies Ltd., Kiryat-Mozkin, Israel

CONTENTS

12.1 Introduction 221
12.2 Technical Characteristics of the Gorona Del Viento Project 222
12.3 Analysis of the Operating Experience of the Gorona Del Viento Project 223
12.4 Analysis of Existing Technologies For Energy Storage and Load Balancing in the Electrical Network 231
12.5 Virtual Power Plant 234
12.6 Conclusions 241
References 242

12.1 INTRODUCTION

The Gorona del Viento hybrid power plant project, implemented in 2014 on the island of El Hierro, is a successful example of the use of renewable energy sources to supply electricity to many consumers.

El Hierro is the smallest of the seven main islands of the Canary archipelago in the Atlantic Ocean and belongs to Spain. The island is home to about 11 thousand people. The area of the island is 268.7 km^2. The main activities on the island are agriculture, animal husbandry and fishing. The island is supplied with power

DOI: 10.1201/9781003257202-13

222   Virtual Power Plant Solution for Future Smart Energy Communities

from its own isolated power plant. There are no connecting cables to other islands. Naturally, all energy carriers are delivered to the island by sea transport. Back in 1997, the Island Council approved a Sustainable Development Plan with the aim of reducing dependence on the outside world and ensuring that the basic needs of El Hierro's residents are met by implementing a life cycle of its own resources that reduce the additional costs associated with living on a small island. One of the main elements of this plan is the supply of electricity to the island using wind power. In fact, a full-scale laboratory has been created on the island of El Hierro for the possibility of using renewable energy to fully meet all the needs of the population. This energy complex includes all the main elements: wind generators, a powerful electric energy storage system and a backup power plant. In the near future, it is planned to additionally build an additional solar power plant.

The operating experience of this hybrid power plant is quite important for the further development of such projects. It should be especially noted that the parameters of all elements of this hybrid power plant are publicly available on the Red Eléctrica de España website in real time, which allows for a detailed analysis of operating experience, as well as to check the effectiveness of using various technologies for active control of an isolated power system to optimize its parameters and improving its efficiency. It is for this reason that the proposed virtual power plant (VPP) project was built in relation to the parameters of the Gorona del Viento hybrid power plant.

Today, the share of renewable energy sources in the total electricity generation balance achieved on the island of El Hierro is almost double that of mainland Europe. In some months, almost all of the electrical energy required is produced using wind power. This is a very high figure. However, the problems currently facing the Gorona del Viento power plant will be inherent in other areas that have not yet reached the same level of renewable energy. Therefore, the operating experience of the Gorona del Viento hybrid power plant is so important to analyze in detail and ensure that the most efficient technologies are applied.

12.2 TECHNICAL CHARACTERISTICS OF THE GORONA DEL VIENTO PROJECT

The hybrid power plant "Gorona del Viento" is located on the island of El Hierro with a population of about 10,000 people. The hybrid power plant includes five wind generators with a total capacity of 11.5 MW, a pumping station for accumulating excess energy with a total capacity of 6.44 MW, four Pelton hydraulic turbines with a total capacity of 11.3 MW, two reservoirs for storing water reserves, the upper one with a capacity of 380,000 cubic meters and the lower one with a capacity of 149,000 cubic meters. The functional diagram of the Gorona del Viento hybrid power plant on the island of El Hierro is presented in Figure 12.1.

When the power generation of electrical energy by wind generators exceeds the total consumption, the pumps pump water from the lower reservoir to the upper one. Thus, excess electrical energy is accumulated. If additional generation power is required, water from the upper reservoir is supplied to Pelton's hydraulic turbines.

Experience of Building a Virtual Power Plant of the Island of El Hierro 223

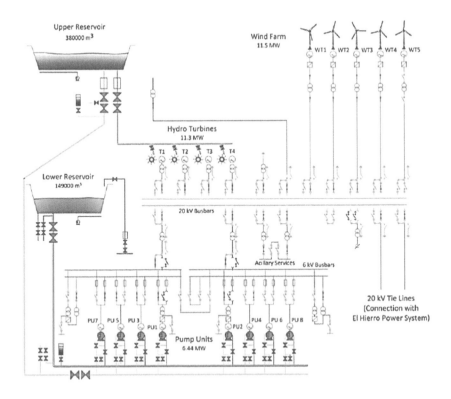

FIGURE 12.1 Functional diagram of the Corona del Viento hybrid power plant [1].

In the absence of wind or lack of generation using renewable energy sources, power supply is carried out using ten diesel generators with a total capacity of 11 MW.

The average daily electricity consumption on the island of El Hierro is 5–6 MW, and the peak load reaches 7–8 MW.

12.3 ANALYSIS OF THE OPERATING EXPERIENCE OF THE GORONA DEL VIENTO PROJECT

The Gorona del Viento hybrid power plant project has always received a lot of attention in the scientific literature from the very beginning of its operation. At the same time, the authors of the studies were specialists from different countries.

Grazyna Jastrzebska from Poznan University of Technology was one of the first to analyze the effectiveness of the Gorona del Viento project in 2018 [2]. But then it was only the initial experimental period of the project operation. In the same year, the creators of the project analyzed the first years of operation of the hybrid power plant and proposed recommendations to improve its efficiency [1]. As will be shown below, only by the end of 2018, this project began to operate steadily at the maximum power level.

A serious analysis of the project's operating experience was presented in a report at the conference "The role of water to produce energy in the Canary Islands", which was held at the University of La Laguna in April 2019 [3]. In this study, for the first time, an analysis was made of the influence of significant fluctuations in the power of electric generation using wind energy. However, the purpose of this study was to ensure the quality of electrical energy and not to increase the efficiency of the entire energy complex. Similar studies are described in the work [4].

Every year, Gorona del Viento El Hierro, S.A., the owner of this project, publishes reports on the operation of the hybrid power plant, which summarizes the performance of all elements of the energy complex [5–7]. However, the main purpose of these reports is to showcase advances in the use of renewable energy sources to power the island.

The work of a group of employees of the Universidad de Las Palmas de Gran Canaria is devoted to the analysis of the efficiency of the Gorona del Viento hybrid power plant [8]. The purpose of this study was to determine and analyze the economic indicators of the project. The technical aspects were not analyzed in this work.

Over the past three years, the island of El Hierro has achieved quite an impressive result of using renewable energy sources for energy supply to consumers. The average over three years share of the use of renewable energy sources in the total balance of electricity production is more than 56%. This is a high result, significantly higher than the average for mainland Europe. In some months of the year, electricity supply to consumers is almost entirely provided by renewable energy sources. Figure 12.2 presents the results of the operation of a hybrid power plant in terms of the use of renewable energy sources.

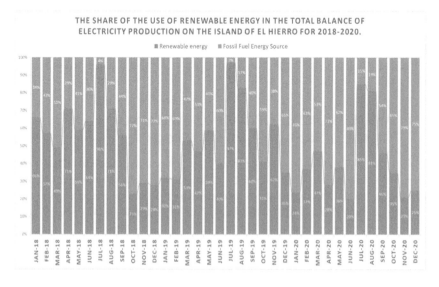

FIGURE 12.2 Share of use of renewable energy sources by months for three years. (From official reports [5–7].)

Experience of Building a Virtual Power Plant of the Island of El Hierro

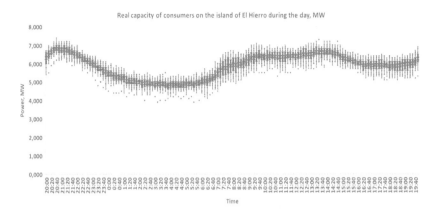

FIGURE 12.3 Schedule of statistical analysis of the power of consumers during the day.

Such a high result has become possible as a result of the hard work of the company's employees and several research institutes in a number of countries. The project is constantly being improved to improve its efficiency.

To date, the planning of the operation of a hybrid power plant is carried out as follows. Consumer power data is constantly analyzed and processed using the apparatus of mathematical statistics. Figure 12.3 presents the results of statistical analysis of real data on the total capacity of consumers on the island of El Hierro during the day.

Based on the data of statistical analysis, a forecast of the power of consumers for the next day is made. Then, based on this data, the optimal operating modes of the generators are planned. With only diesel generators operating, the optimum operating schedule is maintained to ensure maximum plant efficiency. When using renewable energy sources, it is much more difficult to maintain the optimal regime. As a result, there is a deviation from the planned mode of operation.

As shown in Figure 12.4, the real power of consumers is in good agreement with the forecast and planned operation of the generators. It is not always possible to maintain the optimal operating conditions of generators due to the influence of renewable energy sources.

The main problem of all hybrid power plants with a high proportion of the use of renewable energy sources is the constant sharp change in the generation power over time, especially with regard to the use of solar and wind energy. To ensure the reliability of power supply to consumers, special devices and systems are required to maintain a balance in the electrical network. Most often, various technologies for accumulating excess electrical energy or backup sources of electrical energy are used for these purposes. Moreover, most often, the operation of these devices for a long time at partial loads significantly reduces the economic efficiency of the entire energy system.

From the outset, the Gorona del Viento project envisaged balancing the load in the power grid in two ways: by means of a pumped storage system for excess energy and by means of standby diesel generators. Both methods provide sufficient reliability of power supply but have low efficiency and, as a result, lead to large losses of electrical energy.

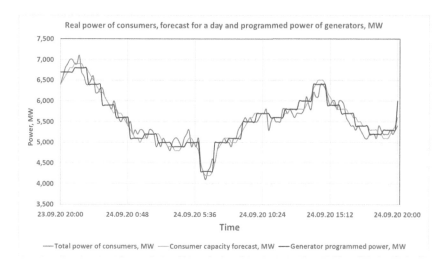

FIGURE 12.4 Real power of consumers, forecast for a day and programmed power of generators, MW

If the pumps and the hydraulic turbine worked for a long time stably at maximum power, then according to the passport data of the equipment, the efficiency of the pumped storage system would be at least 70%. But conditions of constant changes in equipment power, as well as frequent starts and stops, lead to a significant drop in efficiency.

As can be seen in Figure 12.5, in the first three years of operation of the hybrid power plant, load balancing in the electrical grid was carried out mainly with the help of diesel generators. In the period 2015–2017, diesel generators have been operating

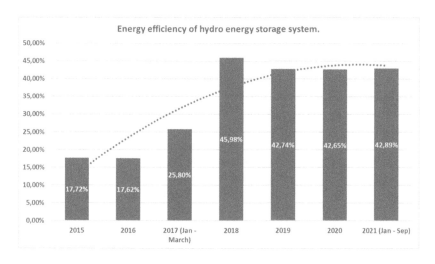

FIGURE 12.5 The efficiency of the system for pumping excess electrical energy.

almost continuously in order to ensure the reliability of electricity supply to consumers. However, this led to an artificial increase in excess electrical energy, and, as a result, the efficiency of the energy storage system was extremely low. Electricity losses in these years amounted to more than 25 GWh per year, which is almost 70% of all electricity produced using wind energy.

Starting in 2018, diesel generators began to be switched off and load balancing began to be carried out using a pumped storage system. Diesel generators were only used during periods of wind power shortages and to prevent emergency situations. This made it possible to increase the efficiency of the energy storage system up to 42–45% and reduce the losses of electrical energy to the level of 10 GWh.

During this period, the operating mode of diesel generators changed. Previously, diesel generators were operated at constant load. Starting in 2018, diesel generators began to operate in the mode of a reserve source of energy supply, which naturally led to an increase in specific fuel consumption. Figure 12.6 shows data on specific emissions of CO_2 into the atmosphere from diesel engines, which are directly proportional to specific fuel consumption.

Any mechanism works with maximum efficiency only at the optimum power level. Any deviation from the optimal operating mode leads to a decrease in efficiency. This applies to all the basic elements of the Gorona del Viento hybrid power plant.

As can be seen from the graphs in Figures 12.7–12.9, all the main elements of a hybrid power plant operate for a sufficiently long time at power levels with low efficiency. This is due precisely to the need to regulate the load in the electrical network using an energy storage system. It should also be noted that the short-term operation of any mechanism is also associated with additional energy losses during the start-up period and during the shutdown period. The shorter the duration of the mechanism, the greater these energy losses.

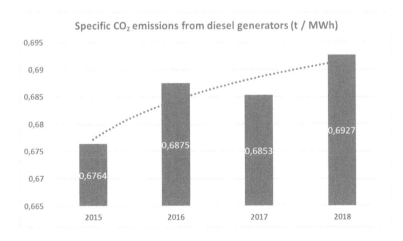

FIGURE 12.6 Specific CO_2 emissions into the atmosphere by diesel generators [9].

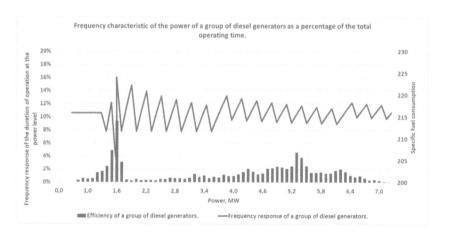

FIGURE 12.7 The efficiency of diesel generators in relation to the power level and the frequency response of continuous operation at a certain power level.

In 2020, in order to ensure the reliability of power supply and the required quality of electrical energy, a storage battery with a capacity of 650 kW and a capacity of 3 MWh was installed at the power plant [3].

This modernization of the project, of course, made it possible to improve the quality of electrical energy in terms of ensuring the specified parameters of the frequency of the electric current, but as can be seen in Figure 12.5, the efficiency of the entire power system remained at the same level.

It should also be noted that the use of energy storage systems for balancing the load in the electrical network leads to an increase in the cost of electricity by two to three times. Only in the case of using the hydroaccumulation technology, the

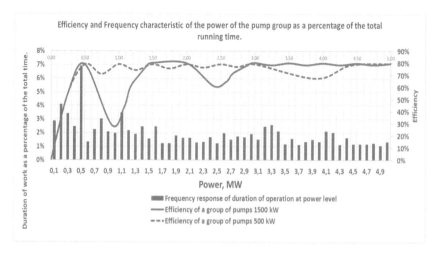

FIGURE 12.8 The efficiency of the two groups of pumps depending on the power level and the frequency response of continuous operation at a certain power level.

Experience of Building a Virtual Power Plant of the Island of El Hierro

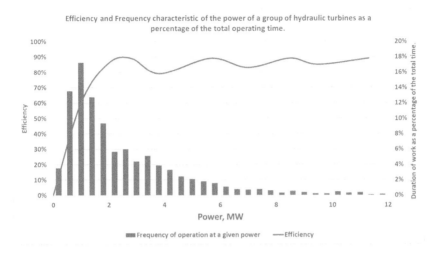

FIGURE 12.9 The efficiency of a group of hydraulic turbines depending on the power level and the frequency response of long-term operation at a certain power level.

increase in cost occurs as a result of low efficiency and large losses of electrical energy, and in the case of using rechargeable batteries, the reason for the increase in the cost of electrical energy is the high cost of the battery.

Also, the disadvantages of hybrid power plants with a high proportion of the use of renewable energy sources should be attributed to a very low utilization rate of power equipment. When on the island of El Hierro, the electricity supply was carried out only with the help of diesel generators, the utilization rate of power equipment was at least 40–50%. Then wind generators with a capacity of 11.5 MW, hydraulic turbines with a total capacity of 11.3 MW and pumps with a capacity of 6.44 MW were installed. As a result, the utilization rate of power equipment fell to 5–6%. It also significantly affects the cost of electricity.

Figure 12.10 shows the power graphs of the main elements of the Gorona del Viento hybrid power plant during October 2020. This figure clearly shows how dramatically the power of the renewable energy source changes over time, which ultimately leads to the same sharp fluctuations in the power of all other elements of the power plant.

Figure 12.11 shows the imbalance between the power of consumers and the power of all generators of electrical energy of the hybrid power plant "Gorona del Viento". And also, for the convenience of the analysis, in Figure 12.11, the power of wind generators and that of diesel generators are duplicated from Figure 12.10.

In Figure 12.11, there are three most characteristic zones. In zone 1, load balancing in the electrical network is carried out only with the help of diesel generators, and the power of wind generators is zero or insignificant. There are no load balancing problems in this zone. Diesel generators have fairly good maneuverability and can cope with fluctuations in consumer power. The power imbalance in this zone is kept constant and equal to 200 kW. This excess power is required for the diesel generator's own needs. In zone 2, the load balancing in the electrical network is carried

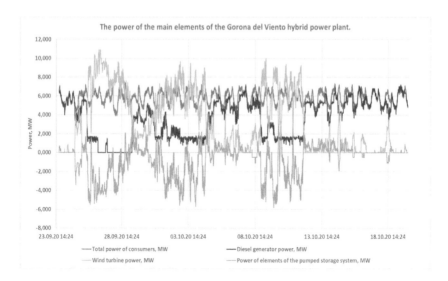

FIGURE 12.10 The capacity of the main elements of the Gorona del Viento hybrid power plant for one month.

out only by the hydraulic energy storage system, and the diesel generators are completely turned off. The imbalance in this zone is significantly higher and has sharp jumps both in the direction of an excess of electrical energy and in the direction of it lack. In zone 3, load balancing in the electrical network is carried out both by diesel generators and a pumped storage system. However, the imbalance and fluctuations over time in this zone are higher than in zone 2.

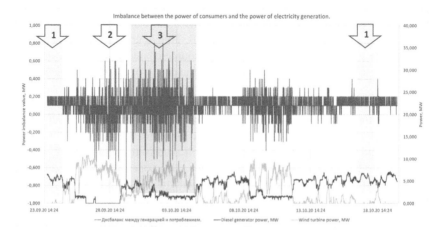

FIGURE 12.11 Imbalance between the power of consumers and the power of the generators of electrical energy of the hybrid power plant "Gorona del Viento".

Experience of Building a Virtual Power Plant of the Island of El Hierro

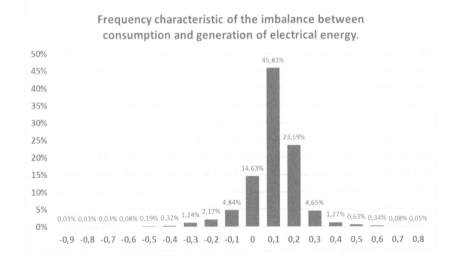

FIGURE 12.12 Frequency characteristic of the imbalance between consumer power and generation power.

Figure 12.12 presents the frequency response of the imbalance between the total power of consumers and the total power of generation. This allows you to estimate possible deviations in the power balance.

A very important parameter for the management of an energy system is the rate of increase and rate of decrease in the power of consumers. The problem is that by no means always standby generators can compensate for the imbalance between the power of consumers and the power of generation rather quickly. This is especially true in the case of a sharp drop in the power of wind generators. This factor should be considered in the design of standby generators and in the design of control systems. The value of this parameter can be obtained only as a result of operating experience with a hybrid power plant. Figure 12.13 shows the data that were obtained based on the analysis of the operating experience of the power plant "Gorona del Viento".

The main purpose of this analysis is to determine quantitative estimates of the possible imbalance in the electrical network with a large share of the use of renewable energy sources, as well as to determine the required volume of energy storage systems to achieve 100% use of renewable energy sources. This allows in the future to evaluate various technologies for energy storage and load balancing in the electrical network.

12.4 ANALYSIS OF EXISTING TECHNOLOGIES FOR ENERGY STORAGE AND LOAD BALANCING IN THE ELECTRICAL NETWORK

At the initial stage of the development of renewable energy sources, the concept dominated those fluctuations in power generation using renewable energy sources would be compensated by the use of energy storage technologies. It was on this

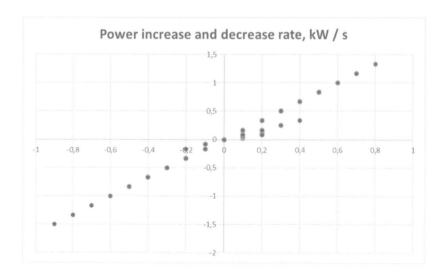

FIGURE 12.13 Possible rate of increase and decrease of the imbalance between generation power and consumer power.

principle that the first large renewable energy projects were designed. However, as practice has shown, this principle is not entirely correct, especially in the case of a high share of the use of renewable energy sources in the total balance of electricity generation.

The main task of energy storage systems is to store a large amount of electrical energy (directly or indirectly) and store energy for a sufficiently long time. It is simply inefficient to use such large power systems for short-term compensation of small energy deficits and inevitably involves large losses of electrical energy.

The main task of load balancing systems in the electrical network is to compensate for short-term deviations from the specified operating mode of the electrical network.

Of course, in either case, energy storage technologies can be used, but the two systems differ significantly in size and energy storage volumes. Therefore, different criteria will be used to analyze and evaluate the effectiveness of a particular technology. For energy storage systems, the main indicators are the volume and duration of energy storage, as well as efficiency. For load balancing systems in the electrical network, the main criteria are agility and, of course, efficiency. At the same time, in this work, only the most common technologies for energy storage and load balancing in the electrical network, which are currently widely used, will be analyzed.

Pumped storage technology is the most widely used worldwide for energy storage. A fairly large number of works are devoted to the study and improvement of this technology [10–15]. Most authors recognize that this technology allows large amounts of energy to be stored for a long time. It is impossible not to agree with this. However, the use of this same technology for load balancing on an electrical grid is controversial. This is contradicted by the operating experience of the Gorona del Viento hybrid power plant. It is inefficient to use the same equipment for energy

Experience of Building a Virtual Power Plant of the Island of El Hierro 233

storage and load balancing in the electrical network. As shown in Figure 12.11, to balance the load in the electricity grid of El Hierro Island, a capacity of 1 MW is required, while power supply of the island during a period of complete wind power outage requires six to eight times more capacity. One and the same equipment cannot operate efficiently in such a wide range of capacities. In addition, the large length of pipelines that connect the upper and lower tanks creates a certain inertia of the system.

A lot of attention has recently been paid to energy storage technology using compressed air [16–18]. This technology has about the same efficiency as the pumped storage system but has a smaller overall dimensions and higher maneuverability. This technology has great development prospects, especially in the case of underwater storage of compressed air [19]. To significantly reduce the overall dimensions of the tanks, it is advisable to use liquefied air or nitrogen [20, 21]. The latter is safer, especially in the case of long-term storage.

A very large development in recent years has occurred in storage batteries of various types [22]. This area of energy storage is developing at a very fast pace. A lot of different projects have been implemented, including quite large ones. The benefits of this technology are undeniable, especially for mobile devices. However, at this point in time, the cost of batteries is quite high, which hinders their widespread use.

Kinetic flywheels can also be used to balance the load in the electrical network [23] and supercapacitors [24]. However, these technologies can only be used for short-term compensation of load changes in the electrical network.

Of particular note is the technology that has been very actively promoted by Wärtsilä Energy in recent years [25]. According to the specialists of this company, fundamentally different technologies should be used for energy storage and for load balancing in the electrical network. To store energy, use one of the above technologies (pumped storage, compressed air or batteries) and use their FLEXIBLE GAS technology to balance the load (flexible engine power plant running on carbon neutral synthetic fuels produced with renewable energy [25]). But at this stage in the development of this technology, natural gas can be used. According to a study by Wärtsilä Energy, FLEXIBLE GAS 3–5% of the total generation capacity is sufficient to ensure load balancing in the grid with 100% renewable energy sources. But this is subject to the presence of a powerful energy storage system. From a technical point of view, this technology is the most optimal.

All of the above technologies are aimed at solving energy storage problems and balancing the load in the electrical network on the side of generating electricity. But a fundamentally different approach is also possible—solving the same problems on the side of electricity consumption. Recently, quite a lot of new technological solutions have appeared, such as VPPs [26, 27], smart microgrids [28, 29], smart consumers [30, 31], energy communities [32, 33], self-sufficiency of consumers with electric energy through the use of small power plants using renewable energy sources [34], etc. Most of these technologies are currently either under development or in trial operation.

The brief analysis of energy storage technologies and load balancing systems in an electric network with a high share of the use of renewable energy sources presented in this section was made only in order to substantiate the following conclusions:

1. Achieving 100% use of renewable energy sources is possible only through the use of energy storage systems. However, technological losses of energy storage systems and the high cost of the necessary equipment will inevitably lead to an increase in the cost of electricity.
2. Significant fluctuations in the generation capacity using renewable energy sources leads to the need to use fundamentally new technologies for balancing the load in the electrical network, especially when the share of renewable energy sources in the total balance of electricity generation exceeds 40–50%.
3. The use of powerful energy storage systems and load balancing systems in the electrical network in modern energy will inevitably lead to a conflict of interests between energy companies and consumers. The problem is that if energy storage systems and load balancing systems in the electrical network are located on the generation side (as part of large power plants), this will inevitably lead to a significant increase in the cost of electricity for consumers. If energy storage systems are placed on the consumption side, firstly, this will lead to an increase in the cost of the necessary equipment of consumers, and secondly, individual consumers of electrical energy will never assume the functions of ensuring the quality of electrical energy in the network and the functions of ensuring balancing load in the network. In order for consumers to take on these functions, they must have some financial preferences, and this will ultimately lead to a decrease in the revenues of energy companies.

The last conclusion is the most important, and the first two only substantiate the inevitability of the third conclusion for modern energy with a high proportion of the use of renewable energy sources.

The problem of conflict of interest can be resolved only by introducing a third party into the overall power supply system, which would ensure the optimization of the power system operation by using the capabilities of both the manufacturer of the electrical system and the capabilities of consumers. Such a third party could be a VPP.

12.5 VIRTUAL POWER PLANT

The term "Virtual power plant" appeared relatively recently, and therefore, different authors give quite different definitions of this term, depending on the field of application. They agree only on one thing, that the core of the VPP is a computer control system that optimizes the operation of the power system.

The most generalized definition is given in the work [35]: "Virtual power plant (VPP) technology can assemble all controllable distributed energy sources of the energy network to collaborate, integrating their scattered capabilities to serve certain needs of Energy Internet". This work also introduces the term "Energy Internet" as a subsection of the Internet of things.

The most complete presentation of the principles of building VPPs is presented in fundamental works [36] and [37]. For each specific case, VPPs can have a very different architecture, depending on the tasks.

Experience of Building a Virtual Power Plant of the Island of El Hierro   235

The main principle of building a VPP in relation to the Gorona del Viento project was the achievement of maximum efficiency of the entire energy system and the maximum reduction of electrical energy losses. To achieve this goal, the following operating mode was proposed. All elements of the hybrid power plant operate at optimal operating conditions with a certain surplus of electrical energy production. Thus, the imbalance between the generation of electrical energy and its consumption is always positive. Due to this, the cost of electricity produced will be minimal. Information about the presence of excess electrical energy is transmitted to the control center of the VPP for its distribution among additional active consumers.

If we take the averaged data for the Gorona del Viento power plant over the past three years, an average of 18,000 MWh was spent on driving the pumps, and the hydraulic turbines generated an average of 8000 MWh. Consequently, if the efficiency of the energy pumped storage system is increased to 55–60%, this will ensure the supply of at least 4000 MWh of free electricity per year to the VPP. At the same time, the power plant itself will not incur additional costs and additional losses of electricity.

Efficiency gains can be realized through the use of a two-stage energy storage system. In the beginning, the surplus electricity is accumulated in the intermediate energy storage. When the accumulated energy is sufficient for a long-term operation of the pump, it is turned on. Then the pump operates at maximum power until the capacity of the intermediate energy storage device is completely depleted. In this way, we can ensure maximum pump efficiency. Similarly, the turbine can operate at optimal power, without surges. A rechargeable battery can serve as an intermediate energy storage, although this option is not optimal. It is much more efficient to use the turbo drive of the pump. The design of such a turbo pump drive is the subject of a separate study.

When comparing the actual and optimal pump modes, as shown in Figure 12.14, the savings in electrical energy consumption for the pump drive in 3.5 hours amounted to about 150 kWh, even considering the electrical energy losses in the battery. In both cases, the amount of water that was pumped into the upper reservoir was 735 cubic meters. This savings will amount to at least 1 MWh per day.

Likewise, it is powerful to achieve savings in the operation of a hydraulic turbine. Even in the complete absence of renewable energy sources, diesel generators with the same fuel consumption will provide the generation of more electrical energy if they operate only at optimal operating conditions. This is clearly seen from Figure 12.7.

Information about the presence of excess electricity in the network is sent to the control center of the VPP in real time. Furthermore, the VPP must distribute this electrical energy among active consumers. The complexity of this process lies in a certain inertia in the management of active consumers. For reliable control of this process, it is necessary to predict the amount and duration of excess electricity in the network.

As already shown in Section 12.1, the Gorona del Viento power plant widely uses the methodology of long-term and medium-term forecasting of the load in the electrical network for the planning of works. Statistical data on the operating modes of all elements of the power plant are constantly analyzed, and, based on this analysis, a work plan is drawn up, at least for a day. However, to effectively manage active

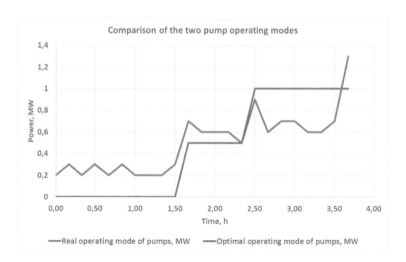

FIGURE 12.14 Comparison of the real and optimal operating mode of a group of pumps with a capacity of 500 kW.

consumers, a short-term forecast of 1 hour or less is required. Without this information, it is practically impossible to organize effective management of the elements of a distributed energy network.

A recent survey report from Guidehouse Inc. states: "As the grid becomes more dynamic and distributed, traditional methods for managing and operating the grid—such as load and generation forecasting, demand response (DR), power quality management, and voltage regulation—are not sufficient. AI solutions enable autonomous and intelligent integration of DER by incorporating real-time data on grid infrastructure, market pricing, supply, demand, load profiles, and weather forecasts" [38].

For short-term forecasting of the load in the electrical network and effective management of active consumers, a universal software package was created using artificial intelligence technologies, neural networks, deep machine learning and the Internet of things, which is the core of a VPP. Figure 12.15 shows a functional diagram of this software package.

Similar studies on the use of deep machine learning using various types of neural networks to predict the load in the electrical network and the power generation of electricity using renewable energy sources, which have been published in recent years, have shown the sufficient effectiveness of this methodology. However, these studies were conducted for short-term predictions of either the intensity of electricity generation from solar panels [39] or for the energy consumption of a separate object [37]. In both cases, we get a forecast of only one component of the energy system, and not the entire system as a whole. To effectively manage the entire energy system, a balance between generation and consumption is required.

As a rule, a neural network is built with many input parameters and one output, which will make it possible to obtain a forecast depending on external parameters. So, for example, to predict the power of a wind generator, the input parameters will

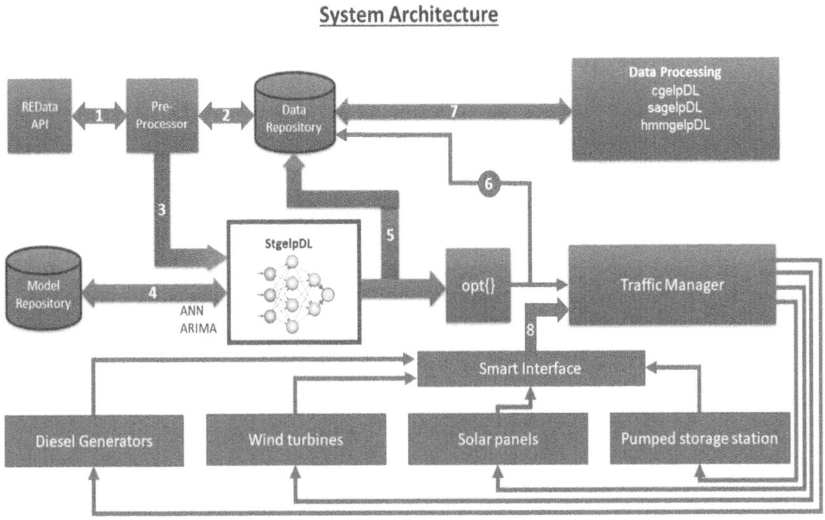

Experience of Building a Virtual Power Plant of the Island of El Hierro

System Architecture

FIGURE 12.15 Functional diagram of the main software complex of the virtual power plant.

be the average wind speed at a given time, the intensity of wind gusts, the frequency of wind gusts, the probability of a sharp drop in wind speed and so on. In the case of using solar energy, the input parameters will be the time of day, the presence and characteristics of cloudiness, the ambient temperature and so on. Such a methodology will require many different sensors, which will constantly measure parameters to obtain a forecast.

Undoubtedly, a large number of factors affect the nature of load fluctuations in the electrical network. These include season, day of the week, climatic parameters and equipment characteristics. However, for short-term forecasting (within 1 hour), all these parameters remain constant or change insignificantly. Therefore, their influence can be neglected. That is why for short-term forecasting, it is possible to use the data of the time series of the output parameter: the power of consumers, the power of each generator or their imbalance.

$$Prediction\ Y_{t+1} = f\left(x_t, x_{t-1} \ldots, x_{t-n}\right)$$

where

- Prediction Y_{t+1}—forecast of a parameter (power of consumers or the amount of imbalance between generation and consumption) for the next time interval.
- x_t—parameter value at a given moment of time.
- x_{t-1}—parameter value at the previous time interval.
- x_{t-n}—value of the parameter n time intervals ago.

It should also be noted that the analysis of the time series of the output parameter makes it possible to consider internal changes in the system and not only changes in external factors.

The software package works as follows. At the first stage, the program reads an array of output parameters of the power system for previous periods through a block through the REData API. Next, the preprocessor processes the received data and forms a matrix for analyzing the time series. The entire data array is divided into two parts. The first part (70%) is used to train neural networks, and the remaining part (30%) is used to test the learning outcomes. The generated data arrays are stored in the Data Repository block. The Predictor then trains and validates the models. At the same time, Predictor trains six different neural networks and two autoregressive models at the same time. This allows in the future to determine the most effective model for forecasting, depending on the application. The resulting models (structures, weights etc.) are stored in the Model Repository block.

In the future, the program reads in real time the current parameters of the energy system, and using the models obtained earlier, a forecast is made several steps ahead. For the Gorona del Viento power plant, the forecast is 10, 20 and 30 minutes ahead. These forecasts are used to optimally manage the energy system of the Gorona del Viento power plant using the Traffic Manager block. After six consecutive predictions, the current data is added to the training sample and the model training process is repeated. Thus, in the process of operation, the forecast models are constantly updated and adjusted.

In addition to the main modules for forecasting the time series of the load in the electric network, the software complex includes additional programs for the analysis and classification of individual time series. These programs help you analyze and make adjustments on the fly.

To check the developed software, real data of the load in the electrical network was used, which was read in real time from the database of the ***Red Eléctrica de España***. The results obtained show a good result, which is shown in Figure 12.16.

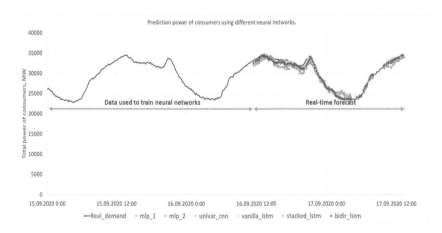

FIGURE 12.16 Results of forecasting the load in the electrical network in real time using six different models.

Experience of Building a Virtual Power Plant of the Island of El Hierro

Separately, a simulation model of the Gorona del Viento hybrid power plant was also created using the SciLab software package (digital twin). With the help of this simulation model, various modes of Traffic Manager operation are configured and the search for optimal solutions for the effective management of all elements of the power plant is done. For example, we take real data shown in Figure 12.10 and make calculations using a simulation model. In the first option, we will use real data for this period to confirm the operability of the simulation model, and in the second option, we will optimize the operation of a group of pumps and diesel generators in accordance with the data presented in Figures 12.7–12.9 and 12.14 by connecting the Traffic Manager block.

In the first version, we get a similar schedule of the elements of the hybrid power plant shown in Figure 12.10, which confirms the correctness of the model. In the second variant, with the help of the Traffic Manager block, pumps and diesel generators operate only at optimal operating modes. The simulation results are shown in Figure 12.17. As you can see in this figure, especially in the highlighted enlarged fragments, diesel generators operate with some excess power but with a sharp transition from one optimal mode to another. Pumps also switch from one optimal mode of operation to another in an abrupt manner, but unlike diesel generators with a lack of use of excess energy. Naturally, in this case, we will have a constant surplus of generated power. It is this excess capacity that must be used by the VPP. The imbalance between generation power and consumer power for the second calculation option is shown in Figure 12.18.

As shown in Figure 12.18, the excess power generated by the hybrid power plant is intermittent. However, by smoothing out power surges using simple special devices, this energy can be successfully used to operate a VPP, at least in part. Moreover, the

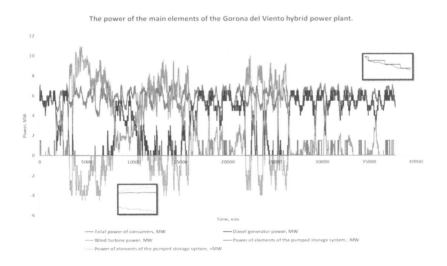

FIGURE 12.17 Results of modeling the operation of the hybrid power plant at the optimal mode.

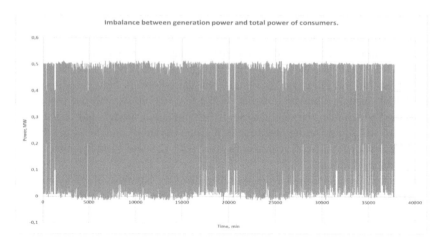

FIGURE 12.18 Imbalance between the generation power and the power of consumers when optimizing the operation of a group of pumps and diesel generators.

presence of a short-term forecast obtained with the help of the developed software package using the artificial Internet will significantly simplify this task.

As shown in Figure 12.19, calculations using a simulation model (digital twin) showed that in both variants, the total energy of consumers and generation using wind energy completely coincide. But as a result of optimizing the operation of a group of pumps and diesel generators, it was possible to reduce the cost of generating electricity using diesel generators by 7%. Moreover, the average specific fuel consumption also decreased by 1.5%. The total effect of reducing CO_2 emissions into

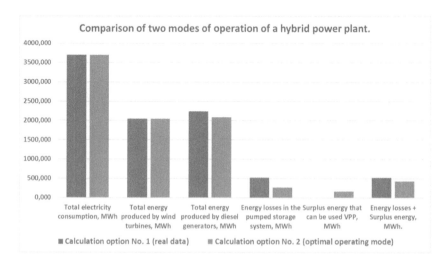

FIGURE 12.19 Comparison of the results of modeling the operation of a hybrid power plant in different modes, real and optimal.

Experience of Building a Virtual Power Plant of the Island of El Hierro   **241**

the atmosphere decreased by 10%. Also, as a result of optimizing the operation of the elements of the hybrid power plant, it was possible to halve the irretrievable losses of electrical energy. Also, the proposed optimization made it possible to provide additional energy at the same cost, which can be successfully used for the operation of a VPP. As can be seen from Figure 12.11, in real conditions, the imbalance of generation and consumption power has deviations both in the positive and negative directions. As a result, the additional energy is zero. In the optimal operating mode proposed in this work, the imbalance of the generation and consumption power has only a positive component, which made it possible to accumulate 162 MWh of additional energy during this period.

It should be especially noted that if the surplus electricity is not used by the VPP and is also converted into irrecoverable losses, all the same, these total losses will be less than in the first version of the calculation.

A resilient and smart grid will help generate predictable and reliable electricity with real-time demand forecasting, operational transparency and analytics across the grid [40].

12.6   CONCLUSIONS

The studies presented in this work allow us to draw the following conclusions:

1. Hybrid power plants with a high range of renewable energy sources (more than 40%) need additional funds to balance the load in the electrical network and ensure the required quality of electrical energy.
2. The use of powerful long-term energy storage systems to solve the problems associated with load balancing does not allow achieving high efficiency of their work. It is necessary to use either two-stage energy storage systems or additional devices.
3. The use of VPP technology can significantly increase the efficiency of hybrid power plants with a high proportion of renewable sources. This is especially true for isolated energy systems. At a minimum, the use of the technology of VPPs will make it possible to turn a part of irrecoverable technological losses into useful energy.
4. For the successful application of the technology of VPPs as part of large power systems, it is first of all necessary to solve the problem of short-term forecasting of the load in the electric network. Without this, efficient distribution of excess electricity is impossible.
5. The software proposed in this work is quite universal and self-organizing. In the process of work, the system constantly makes changes and corrects the created models for forecasting. This makes it possible to take into account not only changes in external parameters but also internal ones, such as equipment wear, block ramen after repair and project modernization.
6. The fact that six different models are formed at once in the software package using various neural networks allows ensuring the reliability of the software package and increasing its versatility.

REFERENCES

1. A. Marrero, Q. Elías, J. Medina Domínguez, J. M. De León Izquier, R. Corujo De León, P. Santos Arozarena, J. Gil, M. Alberto, C. Quintero and J. González Hernández, "Gorona del Viento Wind-Hydro Power Plant Results, Improvement Actuations and Next Steps".
2. G. Frydrychowicz-Jastrzębska, "El Hierro renewable energy hybrid system: A tough compromise," *Energies*, vol. 11, no. 10, p. 2812, 2018.
3. J. Manuel and V. Feijóo, "Baterías como aliados del agua para producir energía eléctrica más limpia en Canarias," 2019.
4. N. Taveira, J. Palomares, E. Quitmann, A. Castañeda and J. Gil, "The hybrid power plant in El Hierro island: facts and challenges from the wind farm perspective." In *Proc. 3rd International Hybrid Power System Workshop*. May 2018.
5. Gorona del Viento El Hierro, S.A., "00-2018-CHGV18Parte_anual-contrata," 2018. [Online]. Available: https://www.goronadelviento.es/informacion-estadistica-y-datos/.
6. Gorona del Viento El Hierro, S.A, "00-2019-CHGV19Parte_anual-Contrata," 2019. [Online]. Available: https://www.goronadelviento.es/informacion-estadistica-y-datos/.
7. Gorona del Viento El Hierro, S.A, "00-2020-CHGV20Parte_anual-contrata," 2020. [Online]. Available: https://www.goronadelviento.es/informacion-estadistica-y-datos/.
8. F. J. Garcia Latorre, J. J. Quintana and I. de la Nuez, "Technical and economic evaluation of the integration of a wind-hydro system in El Hierro island," *Renewable Energy*, vol. 134, pp. 186–193, 2019.
9. C. D. Llanos Blancos, "Declaración Ambiental Enero-Diciembre 2.018 CD Llanos Blancos".
10. N. M. Rozali, S. W. Alwi, Z. A. Manan, J. Klemes and M. Y. Hassan, "Optimisation of pumped-hydro storage system for hybrid power system using power pinch analysis," *Chemical Engineering Transactions*, vol. 35, pp. 85–90, 2013.
11. M. S. Javed, T. Ma, J. Jurasz and M. Y. Amin, "Solar and wind power generation systems with pumped hydro storage: Review and future perspectives," *Renewable Energy*, vol. 148, pp. 176–192, 2020.
12. H. Zhang, D. Chen, B. Xu, E. Patelli and S. Tolo, "Dynamic analysis of a pumped-storage hydropower plant with random power load," *Mechanical Systems and Signal Processing*, vol. 100, no. 100, pp. 524–533, 2018.
13. M. Akkawi, F. Chaaban and A. A. Ghandour, "Design and Cost Analysis for a Pumped-Hydro Storage Power Plant," 2018. [Online]. Available: http://hbrppublication.com/ojs/index.php/jraee/article/view/358. [Accessed 25 10 2021].
14. X. Xu, W. Hu, D. Cao, Q. Huang, C. Chen and Z. Chen, "Optimized sizing of a standalone PV-wind-hydropower station with pumped-storage installation hybrid energy system," *Renewable Energy*, vol. 147, pp. 1418–1431, 2020.
15. X. Wu, Y. Xu, J. Liu, C. Lv, J. Zhou and Q. Zhang, "Characteristics analysis and fuzzy fractional-order PID parameter optimization for primary frequency modulation of a pumped storage unit based on a multi-objective gravitational search algorithm," *Energies*, vol. 13, no. 1, p. 137, 2019.
16. Y. Li, S. Miao, B. Yin, W. Yang, S. Zhang, X. Luo and J. Wang, "A real-time dispatch model of CAES with considering the part-load characteristics and the power regulation uncertainty," *International Journal of Electrical Power & Energy Systems*, vol. 105, pp. 179–190, 2019.
17. H. Meng, M. Wang, O. Olumayegun, X. Luo and X. Liu, "Process design, operation and economic evaluation of compressed air energy storage (CAES) for wind power through modelling and simulation," *Renewable Energy*, vol. 136, pp. 923–936, 2019.
18. G. Venkataramani, P. Parankusam, V. Ramalingam and J. Wang, "A review on compressed air energy storage – A pathway for smart grid and polygeneration," *Renewable and Sustainable Energy Reviews*, vol. 62, pp. 895–907, 2016.

19. Z. Wang, R. Carriveau, D. S. Ting, W. Xiong and Z. Wang, "A review of marine renewable energy storage," *International Journal of Energy Research*, vol. 43, no. 12, pp. 6108–6150, 2019.
20. I. Lee and F. You, "Systems design and analysis of liquid air energy storage from liquefied natural gas cold energy," *Applied Energy*, vol. 242, pp. 168–180, 2019.
21. H. Qing, W. Lijian, Z. Qian, L. Chang, D. Dongmei and L. Wenyi, "Thermodynamic analysis and optimization of liquefied air energy storage system," *Energy*, vol. 173, pp. 162–173, 2019.
22. A. R. Dehghani-Sanij, E. Tharumalingam, M. B. Dusseault and R. Fraser, "Study of energy storage systems and environmental challenges of batteries," *Renewable and Sustainable Energy Reviews*, vol. 104, pp. 192–208, 2019.
23. X. Dai, K. Wei, X. Zhang, X. Dai, K. Wei and X. Zhang, "Analysis of the peak load leveling mode of a hybrid power system with flywheel energy storage in oil drilling rig," *Energies*, vol. 12, no. 4, p. 606, 2019.
24. A. Stott, M. O. Tas, E. Y. Matsubara, M. G. Masteghin, J. M. Rosolen, R. A. Sporea and S. R. P. Silva, "Exceptional rate-capability from carbon encapsulated polyaniline supercapacitor electrodes," *Energy & Environmental Materials*, vol. 3, pp. 389–397, 2020.
25. W. Energy, "Aligning Stimulus with Energy Transformation – A report by Wärtsilä Energy".
26. E. A. Bhuiyan, M. Z. Hossain, S. M. Muyeen, S. R. Fahim, S. K. Sarker and S. K. Das, "Towards next generation virtual power plant: Technology review and frameworks," *Renewable and Sustainable Energy Reviews*, vol. 150, p. 111358, 2021.
27. S. Morteza Alizadeh, "Introduction to Virtual Power Plants ".2021, [Online]. Available: https://www.eit.edu.au/wp-content/uploads/2021/07/VirtualPowerPlants_15.07.21_WithRecording.pdf.
28. M. H. Christensen, R. Li and P. Pinson, "Demand side management of heat in smart homes: Living-lab," *Energy*, vol. 195, p. 116993, 2020.
29. N. Dkhili, J. Eynard, S. Thil and S. Grieu, "A survey of modelling and smart management tools for power grids with prolific distributed generation," *Sustainable Energy, Grids and Networks*, vol. 21, p. 100284, 2020.
30. G. J. Osório, M. Shafie-Khah, G. C. Carvalho and J. P. Catalão, "Optimal residential community demand response scheduling in smart grid," *Applied Energy*, vol. 210, pp. 1280–1289, 2019.
31. "Nace el proyecto piloto Gamma, un laboratorio demostrador de la gestión digitalizada de la energía • SMARTGRIDSINFO," [Online]. Available: https://www.smartgridsinfo.es/2020/03/06/nace-proyecto-piloto-gamma-laboratorio-demostrador-gestion-digitalizada-energia.
32. N. Vespermann, T. Hamacher and J. Kazempour, "Access economy for storage in energy communities," *IEEE Transactions on Power Systems*, vol. 36, no. 3, pp. 2234–2250, 2021.
33. A. Tartaglia, "Energy communities," *E3S Web of Conferences*, 2019.
34. J. Dehler-Holland, D. Keles, T. Telsnig and M. J. Baumann, "Self-consumption of electricity from renewable sources EMHIRES: European Meteorological-derived High resolution RES dataset View project Solar Mining View project," 2015.
35. W. Su (Redactor) and Q. Huang (Redactor), The Energy Internet: An Open Energy Platform to Transform Legacy Power Systems into Open Innovation and Global Economic, Woodhead Publishing, 2019.
36. C. Ninagawa, Virtual Power Plant System Integration Technology, Springer, Singapore, 2020.
37. L. Baringo and M. Rahimiyan, Virtual Power Plants and Electricity Markets, Springer, Cham, 2020.

38. H. Davis, S. Research, A. Roberto and R. Labastida, "Executive Summary: AI for DER Integration AI for DER Integration Use Cases and Market Trends across Grid Management, Demand Side Management, and Customer-Centric Segments: Global Market Analysis and Forecasts, 3Q, 2021.
39. M. Grogan, "ARIMA vs. LSTM: Forecasting Electricity Consumption," Towards Data Science. [Online]. Available: https://towardsdatascience.com/arima-vs-lstm-forecasting-electricity-consumption-3215b086da77.
40. "Sustainable Grid Solutions | Emerson US," [Online]. Available: https://www.emerson.com/en-us/industries/automation/power-generation/sustainable-grid-solutions?utm_source=emr-s&utm_medium=vtye&utm_content=Emerson.com/WeSeeSustainableGridSolutions&utm_campaign=21gEMR-WeSeeSustainableGridS01#E11.

Index

Note: Locators in *italics* represent figures and **bold** indicate tables in the text.

A

Active distribution networks (ADNs), 13
Advanced distribution management systems (ADMSs), 193
Aggregators, 57–59, 62
Ancillary services (ASs), 63–64
Automatic regressive distributed log (ARDL), 30

B

Battery energy storage system, 136–138
Bitcoin, 145–146
Bitumen tanks (BTs), 136–138
Blockchain techniques, 149–151
  economic feasibility, 153
  energy consumption, production, and exchange, 29–31
  microgrid applications, 28–29
  types, 27–28
Branch-and-bound method, 60
BTs, *see* Bitumen tanks (BTs)

C

CI, *see* Complementarity index (CI)
Closed cycle gas turbine (CCGT), 98
Coefficient of correlation (R), 49
Coefficient of variation (CV), 49
Co-integration analysis, 30
Column generation method, 60
Commercial virtual power plant (CVPP), 14, 16–18, 109
Communication systems, 127
Complementarity index (CI), 210–213
Consensus-based approach, 178–180
Control technique, 47–48
CVPP, *see* Commercial virtual power plant (CVPP)

D

Decentralization, 146
Decentralized applications (DApps), 149
Decentralized energy management system, 175–177
Demand response (DR), 40
  electricity markets, 50
  red state, *73*

Demand response communication protocols, 45–46
DERs, *see* Distributed energy resources (DERs)
Deterministic model, *26*
DG, *see* Distributed generation (DG)
Direct control techniques, 47
Distributed energy resources (DERs), 8, 9, *57*
  demand response applications, *62*
  energy management, 24–25
Distributed energy resources customer adoption model (DER-CAM), 36
Distributed generation (DG), 111, 198–200
Distributed ledger technology (DLT), 65
Distributed network protocol, version 3 (DNP3), 194
Distribution system operator (DSO), 19
  and traffic light concept, 64–65
DLT, *see* Distributed ledger technology (DLT)
Domestic distributed generator (DDG), 13
Dynamic virtual power plants (VPPs), 58–59

E

EC, *see* Energy community (EC) model
Economic optimization, 94–97
Edge-computing strategy, 203–204
Edge-Database architecture, 200
Electricity consumption, 49
EMS, *see* Energy management system (EMS)
End-user energy costs, 97
Energy community (EC) model, 109–110
Energy management system (EMS), 44, 82
  case study, 180–181
  problem formulation, 177–178
  simulation results, 181–188
  VPP, 175
Energy markets
  aggregator, 62–63
  DERs, 62–63
  DSO, 64–65
  prosumers, 62–63
  technologies, 65–66
  traffic light concept, 64–65
  types, 63–64
Energy storage (ES), 111–112
Energy storage systems (ESSs), 80
  conomic impacts, 91–93
  environmental impacts, 91–93

245

246

Index

system design, 84–85
systems modelling, 85–90
Environmental impact assessment, 97–98
Equilibrium trading, 23
ESSs, *see* Energy storage systems (ESSs)

F

First-order discrete-time consensus protocol,
178–180
Flexibility index (FI), 213–215
Flexible loads (FLs), 112
Flywheel energy storage units
central controller, 130–133
local controllers, 130–133
Forecasting methods, 65–66
Formation methodology
algorithm, 67–69
green state, 70
grid states, 66–67
red state
computational complexity, 72–76
use case, 72
yellow state, 70–71
Fuel cell, 88–90

G

Game theory, 60–61
Geographical islands, 82
Gorona del Viento hybrid power plant
characteristics, 222–223
energy storage, 231–234
load balancing, 231–234
operating experience, 223–231
VPP, 234–241

H

Heuristic methods, 61
HomePlug® Alliance, 46
HOMER Energy, 82
Hydrogen electrolyser, 88–90

I

ICT, *see* Information communication technology
(ICT)
IEC 61850 protocol, 194, 202–203
IEEE 14 bus system, 31–33
impedance, **34**
IEEE 30 bus system, 33–34, *35*
IEEE 1815-2010 protocol, 194
Immutability, 146
Indirect control techniques, 47
Information communication technology
(ICT), 16–18

Integer programming, 25
Internet of Things (IoT), 65
Interoperable VPP
case study, 215–218
complementarity, 210–213
flexibility, 213–215

L

Large-scale VPPs (LSVPPs), *15*
LCA, *see* Life cycle assessment (LCA)
LEM, *see* Local energy market (LEM)
Levelized cost of electricity (LCOE), 92
LFM, *see* Local flexibility market (LFM)
Life cycle assessment (LCA), 93
Linear models, 25
Linear programming, 25
Lithium-ion battery, 87–88
Local energy market (LEM), 63
Local flexibility market (LFM), 63
Local flexibility market operator (LFMO), 70
Local VPPs (LVPPs), *15*
Logic-based energy management system (EMS),
47
Logic control system, 90–91
Low-voltage (LV) microgrid, 22–23
LSVPPs, *see* Large-scale VPPs (LSVPPs)
LVPPs, *see* Local VPPs (LVPPs)

M

MAPE, *see* Mean absolute percentage error
(MAPE)
Market aggregators, 19
Mean absolute error (MAE), 49
Mean absolute percentage error (MAPE), 49
Mean square error (MSE), 49
Medium-voltage/low-voltage (MV/LV)
transformer, 23
Microgrid, **11**
Microgrid basic model, *192*
Micro-PMU (MPMU), 200–202
Microrequest-based aggregation, forecasting, and
scheduling of energy demand, supply,
and distribution (MIRABEL), 46
Mixed-integer linear programming (MILP), 25,
156
Modelling techniques, 82
Model predictive control (MPC), 47
MSE, *see* Mean square error (MSE)
Multiobjective optimization, 61

N

Network time protocol (NTP), 195–196
Non-linear models, 25
Numerical analyses, 112–116

Index

247

O

Opal-RT technologies, 195
Open automated demand response (OpenADR), 46
Open field message bus (OpenFMB), 193
Optimization methods, 24–25, 66, 82

P

PDG, *see* Public distributed generator (PDG)
Peer–peer energy (P2P) trading
  design, 20–22
  integrated distributed grid control, 20
  LV microgrid, 22–23
PEMFC, *see* Proton exchange membrane fuel cell (PEMFC)
Photovoltaic solar model, 85–87
PLC, *see* Programmable logic controller (PLC)
Power purchase agreements (PPAs), 64
Power quality (PQ), **17**
PPAs, *see* Power purchase agreements (PPAs)
Precision time protocol (PTP), 195–196
Price-based control mechanism, 63
Probabilistic model, *26*
Programmable logic controller (PLC), 45
Prosumers, 6263
Proton exchange membrane fuel cell (PEMFC), 88
PTP, *see* Precision time protocol (PTP)
Public distributed generator (PDG), 13
Pumped storage technology, 232

R

Real-time distributed clustering approach, 41
Real-time operating systems (RTOSs), 197–198
Real-time PTP architecture, 198–200
Refrigerators
  central controller, 130–133
  local controllers, 130–133
Regional VPPs (RVPPs), *15*
Regulators, 19
Renewable energy-based microgrid, 41–43
Renewable energy systems (RES), 80
Renewable intermittency, *9*
RMSE, *see* Root mean square error (RMSE)
Root mean square error (RMSE), 49

S

Scalability, 22
Smart building management technologies, 127
Smart contracts, 148–149

Smart grids, **11**, 43–46
Smart switches, 126

T

Technical virtual power plant (TVPP), 16, 105–109
Time-sensitive networking (TSN), 195–196
TLC, *see* Traffic light concept (TLC)
Traceability, 146
Traditional grid, **11**
Traffic light concept (TLC), 64–65
Transmission system operator (TSO), 19
TSN, *see* Time-sensitive networking (TSN)
TSO, *see* Transmission system operator (TSO)
TVPP, *see* Technical virtual power plant (TVPP)

V

Virtual energy storage system (VESS), 120–121
  applications, 127–128
  benefits, 128–129
  control scheme, 129–130
  ESS, 122–126
  flexible demand units, 121–122
  frequency control scheme
    case study, 133–134
    central controller, 130–133
    components, 130
    local controllers, 130–133
  technologies, 126–127
  voltage control scheme
    case study, 138–141
    components, 135
    distributed controllers, 136–138
Virtual power plants (VPPs), 8–9, **11**
  advantages, 9–10
  aggregation, 59
  case study, 15–18
  disadvantages, 10
  electricity markets and networks, *12*
  elements, 10–11, *12*
  emerging technologies, 20–25
  EMS, 174–175
  energy storage technology, 13–14
  ESS, 81–82
  formation (*see* Formation methodology)
  generic benefits, 18–19
  Gorona del Viento hybrid power plant, 234–241

implementation, 44–45
new regulations in India, 36–38
objectives and solving approaches, 60–61
research trend, **17**
technology generation, 11–13
types, 14–15
VPP aggregator (VPPA)
DA energy markets price, *166*
decision-making framework, 158–159
mathematical formulation, 163–166
reserve market price, *167*
RT energy markets price, *166*
strategies, 159–160

W

Wholesale (WS) markets, 64

Z

ZigBee, 46